1985

Pollution of our Atmosphere

Pollution of our Atmosphere

Dr B Henderson-Sellers
University of Salford

Adam Hilger Ltd, Bristol

British Library Cataloguing in Publication Data

Henderson-Sellers, B.
   Pollution of our atmosphere.
   1. Air—Pollution
   I. Title
   363.7'392   TD883

   ISBN 0-85274-754-3 (hbk)
   ISBN 0-85274-763-2 (pbk)

Consultant Editors: **Professor J M A Lenihan**, Western Regional Hospital Board, Glasgow and **Dr G Nickless**, University of Bristol

Published by Adam Hilger Ltd, Techno House, Redcliffe Way, Bristol BS1 6NX

Typeset by Mathematical Composition Setters Ltd, Salisbury, UK and printed in Great Britain by J W Arrowsmith Ltd, Bristol

To Ann

Contents

Preface

Pollutant emissions to the atmosphere change temporally and spatially as a result of technology, social attitudes, population movement and increase or decrease, economic pressure, fuel usage etc. The existence of an air pollution problem results largely from public perception of the environment and people's concern that existing technology could be used to improve the air quality of their locale. There is thus no absolute quantitative definition of air pollution—both legal standards and socially acceptable concentrations depend upon the political and socioeconomic structure of a country and its degree of industrial development. Indeed it is often observed that during the early stages of industrialisation, economic pressures completely overwhelm environmental concerns. Pollutant control can be costly, and only when all industries feel themselves to be economically established can they afford the time and money to consider implementation of controls over their waste products (with the exception of those industries where the wastes contain economically recoverable constituents). Within this scenario, we can consider that guidelines for pollution have already been established by the industrialised world during centuries of development but adoption of these 'standards' by the Third World remains an open question. The examples given are, largely of necessity (due to the greater availability of data on all aspects of air pollution and pollution control), limited to Europe and North America. However air pollution knows no political or geographical boundaries; long range transport of sulphur dioxide, for example, leads to international exchange of pollutants as typified by the acid rain problems of Canada and Scandinavia. In all countries (developed or developing) the chemistry and physics remain the same. In many cases too the technology is immediately transferable between countries. Hence the use of a restricted data set in no way restricts the applicability of the conclusions. To further the international relevance of this text SI units are used throughout (in the author's view a necessity in modern science).

The material in this volume is based largely on a series of lectures given at the University of Salford to a group of students studying environmental science with special emphasis on the health effects of pollution within the urban environment. It is almost impossible for any student entering upon this course of study to possess background knowledge of sufficient depth to encompass the many aspects of this subject without first the whole class being directed to revision or review material in order to establish a common base. The choice of this I leave to the individual tutor or course leader. The material in this book can also be (and has been) used in courses on Environmental Chemical Engineering and Environmental Resources at final year Honours and Masters degree level, but in this case the text supplies basic information which is then supplemented by reference to papers from the current research literature.

The text follows a structure which, it is hoped, will be found useful for a wide variety of atmospheric pollution courses. It covers all aspects of the subject: physics, observational techniques, health effects, fuels and industrial combustion, prevention, control and legislation—all subjects about which the environmental scientist dare not be ignorant, even if his expertise is highly specific. Chapter 1 describes the historical development of pollution and the chemical pollutants of interest. Some emissions figures are also given. Chapter 2 introduces basic air pollution meteorology and climatology. Chapter 3 looks at the problem of stack emissions (the predominant industrial source). This includes quantitative analysis of diffusion, plume rise and pollutant removal. Chapter 4 then describes the methods of assessing pollutant concentrations, together with some recent results for ambient values. The next link, from emissions through transport to removal and measurements, is described in Chapter 5 in which the deleterious effects on plants, animals and man are considered in detail. Chapter 6 then describes pollutant creation (i.e. before emission) in terms of fuels, combustion chemistry, furnaces and boilers. Chapter 7 deals with the modern day problem of mobile sources including photochemical smog formation and Chapter 8 with the technology of pollution control (including a section on the legislative requirements in different countries).

This text could not have been completed without the assistance and encouragement of many people: colleagues at both Salford University and Woods Hole Oceanographic Institution where the text was completed and also student groups over the years. Specific thanks go to Professor P Slawson, Dr P Brimblecombe, Dr D Trout, Dr D S Munro and Dr J W Bacon for reading the draft manuscript. I wish to thank Alan Abbott for the cover photograph and figure 3.15; Bert Pade for figure 1.8; Sara Whiteley for figure 1.13; Kendal McGuffie for figures 3.1 and 3.2; Tom Lyons for figure 3.3 (lower); Alex Alexander for figure 4.4 (right) and my brother Peter for figure 3.12 (left).

Finally thanks to my wife Ann for her love and patient understanding without which this project could never have been completed.

Brian Henderson-Sellers
Louvain-la-Neuve, April 1983

A Note on SI Units

Perhaps the most important aspect of a quantitative approach is the correct and sensible use of a units system. In modern science the SI system (Système International) is advocated. In this system the basic units of interest† are:

length	metre (m)
mass	kilogram (kg)
temperature	kelvin (K)
time	second (s)
amount of substance	mole (mol).

Three useful derived units are:

force	newton (N) ($1 \text{ N} = 1 \text{ kg m s}^{-2}$)
energy	joule (J) ($1 \text{ J} = 1 \text{ kg m}^2 \text{s}^{-2} = 0.2388 \text{ cal}$)
power	watt (W) ($1 \text{ W} = 1 \text{ J s}^{-1}$).

Perhaps the two most used units in air pollution literature are those of length and concentration (amount of substance in a given volume). Although lengths (in the SI) should be given in metres, many of the sizes encountered (e.g. sizes of respirable particulates) would require a large negative power of ten and hence smaller units (in multiples of $10^{\pm 3}$) are permitted in the system. Thus

$$10^6 \, \mu\text{m (micrometre)} = 10^3 \text{ mm (millimetre)} = 1 \text{ m.}$$

Similarly for mass

$$10^6 \, \mu\text{g} = 10^3 \text{ mg} = 1 \text{ g} = 10^{-3} \text{ kg}$$

i.e. $\quad 10^9 \, \mu\text{g} = 10^6 \text{ mg} = 10^3 \text{ g} = 1 \text{ kg.}$

†The other two basic units are electric current (ampere), luminous intensity (candela)

1

Introduction—Background and Source Types

Pollution—the very word is emotive. In the study of pollution of our atmosphere it is necessary to see through this emotion to the facts beneath. A definition of our subject matter is necessary. Air pollution is defined in as many ways as there are authors, for example:

'Air pollution means the presence in the outdoor atmosphere of one or more contaminants, such as dust, fumes, gas, mist, odour, smoke, or vapour in quantities, of characteristics, and of duration, such as to be injurious to human, plant or animal life or to property, or which unreasonably interferes with the comfortable enjoyment of life and property' (quoted in Perkins 1974)

'An unfavourable alteration to the environment' (Hodges 1973)

'The presence in the atmosphere of a substance or substances added directly or indirectly *by an act of man* [author's italics], in such amounts as to affect humans, animals, vegetation, or materials adversely' (Williamson 1973).

Dictionary definitions of the act of polluting are several and include

'to make or render unclean; to defile, desecrate, profane' (*Websters*)

'to make foul; to desecrate; to corrupt' (*Collins*)

'to destroy the purity or outrage the sanctity of' (*Shorter Oxford*).

   Taking an overview of these definitions leads to two fundamental questions. If air pollution is defined, in some way, as the addition of undesirable substances to the atmosphere, then we need to know the composition of the 'undefiled' atmosphere. Although it is a moot point as to what is natural it could be argued that only a world without man was unpolluted. The origin and evolution of life itself probably caused the most traumatic chemical change in the atmosphere: the introduction of free oxygen. Indeed the atmospheric composition given in table 1.1 has been determined largely by biological, rather than inorganic, processes. Notwithstanding, it is usual to present data from an unpolluted atmosphere which is based largely on an Earth before industrialisation. The values in table 1.1 are given as a volume ratio (i.e. % or ppm) and show the predominance of gaseous oxygen and the relatively inert gas nitrogen. Also in the table are found the background values for the 'pollutants' such as SO_2, NO_2 and O_3—values in excess of those quoted here indicate local pollution.

Table 1.1   The composition of the dry atmosphere (after Schidlowski 1980).

Constituent	Chemical formula	Abundance by volume
Nitrogen	N_2	78.084 ± 0.004 %
Oxygen	O_2	20.948 ± 0.002 %
Argon	Ar	0.934 ± 0.001 %
Water vapour	H_2O	variable (%−ppm)
Carbon dioxide	CO_2	325 ppm
Neon	Ne	18 ppm
Helium	He	5 ppm
Krypton	Kr	1 ppm
Xenon	Xe	0.08 ppm
Methane	CH_4	2 ppm
Hydrogen	H_2	0.5 ppm
Nitrous oxide	N_2O	0.3 ppm
Carbon monoxide	CO	0.05−0.2 ppm
Ozone	O_3	variable (0.02−10 ppm)
Ammonia	NH_3	4 ppb
Nitrogen dioxide	NO_2	1 ppb
Sulphur dioxide	SO_2	1 ppb
Hydrogen sulphide	H_2S	0.05 ppb

Alternatively, if substances are added to the atmosphere, not by man but by another part of the biosphere or by natural catastrophe, should this be regarded as natural air pollution, in contrast to industrial/domestic or *anthropogenic* air pollution? It is deemed worthwhile to discuss both these types of air pollution. It will be shown, perhaps surprisingly, that natural emissions of many so-called pollutants are in fact much greater than anthropogenic emissions; yet this book will deal almost exclusively with the latter. Why is this justified? To answer this question it is necessary to trace the historical tale.

For the present discussion let us assume *pollution* encompasses natural and man-made sources. Historically natural sources have always been present in the form largely of volcanoes, forest fires, dust storms, sea-salt spray etc. Volcanoes add excess dust and gases such as sulphur dioxide and carbon dioxode to the atmosphere. Forest fires, perhaps initiated by lightning, add large quantities of smoke (carbonaceous particles), unburned hydrocarbons (any chemical compound containing hydrogen and carbon),

carbon monoxide and oxides of nitrogen to the atmosphere. Such fires may be visible for several hundred kilometres. A dust storm is a natural phenomenon, creating large amounts of suspended particulate matter as a natural 'pollutant'.

HISTORICAL ASPECTS

As soon as man first 'tamed' the element of fire, anthropogenic pollution became possible—but with very limited (both spatial and temporal) effects. Indeed polluting one's own immediate environment is considered socially acceptable—it is with increasing population and more 'sophisticated' technology that man pollutes his neighbour's environment and his pollution then becomes a problem.

Fire, or more technically *combustion*, is responsible for over ninety per cent of air pollutants. The domestic fires of our cave-dwelling ancestors developed into congregations of fires in the urban environments of the Roman Empire† and the communal fires of the mediaeval manor and subsequently, industrialised Europe and America. Smoke (and other byproducts of combustion) rose from the fires, making the air in the immediate vicinity 'heavy laden'. The burning of fires indoors lead to the invention of the *chimney*, creating both a draught and an outlet for removing smoke from the place of origin (figure 1.1). Chimneys do nothing for air pollution control (see Chapter 8) but are simply designed to redistribute the pollution out of the ken of the originator. The domestic chimney is not always successful in this aim, being relatively squat, but the industrial chimney is designed to be tall enough to spread its correspondingly greater emission over a large area (at the same time ensuring it dilutes sufficiently to be of little harm—see Chapter 3).

Combustion-generated pollutants are determined by the nature of the fuel (see Chapter 6). In Britain, wood was the prime fuel for both domestic and industrial use until the end of the thirteenth century. The first recorded instance of smoke pollution from coal burning would appear to have been in 1257, when emissions from coal being used in the building works for the renovation of Nottingham Castle were so great that Queen Eleanor (Henry III's wife) was forced to leave Nottingham for the less polluted rural surroundings of Tutbury Castle. By the early fourteenth century the use of coal had started to become common, particularly as an industrial fuel. Although it is thought that coal was first used in Britain around 1500 BC in the Bridgend (Glamorgan) area, it was not until 1180 AD that systematic

---

†In 61AD Seneca reported 'As soon as I had gotten out of the heavy air of Rome and from the stink of the smoky chimneys, thereof, which being stirred, poured forth whatever pestilential vapours and soots they had enclosed in them, I felt an alteration in my disposition.'

mining began and by 1228 coal was being shipped regularly to London from the Newcastle area. Dick Whittington's cat was, in reality, a coal barge! Another source of coal was seafloor coal extrusions which were subject to marine erosion. Coal was deposited on beaches—hence its early name of sea coal—a name commemorated in Seacoals Lane in London.

Figure 1.1 Tudor chimneys, Harefield, Middlesex (UK).

In 1273 Edward I took the first legal steps to alleviate atmospheric pollution by prohibiting coal burning in London. This apparently had no long term effect as an identical proclamation was given by Elizabeth I. At one time (under Edward II) the use of coal carried the death penalty. By the seventeenth century King Charles II was sufficiently concerned about London's smoke (the city today is still referred to familiarly as 'The Smoke') to commission a report by John Evelyn entitled *Fumifugium, or the Inconvenience of the Aer, and Smoake of London Dissipated (together with some remedies humbly proposed)*—a work which is still worth reading and which, happily, was reprinted recently.

Figure 1.2 Melanism in the peppered moth. Photographs by John Haywood, University of Oxford, Department of Zoology.

Before the Industrial Revolution, coal-consuming industries were on a small scale, largely confined to metallurgy, lime burning, ceramics, preservation of animal products, bricks and leather tanning. Coke did not become important in metallurgical processes until 1700. The harnessing of steam power following inventions such as Watt's reciprocating steam engine in 1784 resulted in greatly increased consumption of coal over more widespread geographical areas. Governmental concern began to increase in the late nineteenth centry. The first Public Health Act of 1848 was followed by the Alkali Act of 1863. American legislation was first directed at industrial and locomotive emissions although in both countries a smoking chimney was regarded as a symbol of a successful business! Much US legislation was either municipal or state legislation: in 1867 the city of St Louis demanded that chimneys should be 20 feet (about 6 metres) higher than adjoining buildings. However an Act against smoke passed in 1893 was unable to be enforced until 1902. In Britain, Manchester led the way in air pollution control. Although the Corporation initiated a Nuisance Committee on Smoke in 1801, powers were not given to the police for control of smoke until the Manchester Police Act of 1844. Although some progress was made, the first sighting of the melanic form of the peppered moth (*Biston betularia*) (figure 1.2) occurred in Manchester in 1848. Black smoke emissions did not become an offence until 1866, and even the legislative enforcement of fines of £10 in 1882 were by no means preventative. The concept of smokeless zones was first raised in Manchester by Mr Gandy in 1934—plans for which were delayed until 1946 by the Second World War.

Before the Clean Air Act of 1956, 221.5 hectares of Manchester had already been declared 'smokeless'.

The Ringelmann chart was perhaps the first attempt to quantify emissions. This concern with smoke measurement and smoke abatement was accelerated by the London smog† episode of 1952. London 'pea-soupers' were by that time infamous and accepted as part of everyday life—Dickens referred to them as the London 'familiar'. In *A Christmas Carol*, written in December 1843, he describes a London afternoon: 'The city clocks had only just gone three, but it was quite dark already: it had not been light all day: and candles were flaring in the windows of the neighbouring office, like ruddy smears upon the palpable brown air'. Cattle deaths at the Smithfield show had been attributed to air pollution since 1873. This was compounded during the period December 4–8 1952 by human deaths calculated at 4000 over and above the expected figures (figure 1.3). Other extended pollution periods or *episodes* to note were those at Seraing in the Meuse Valley (3 days in 1930), Donora, Pennsylvania (1948) and the US east coast (Thanksgiving, November 22–25 1966).

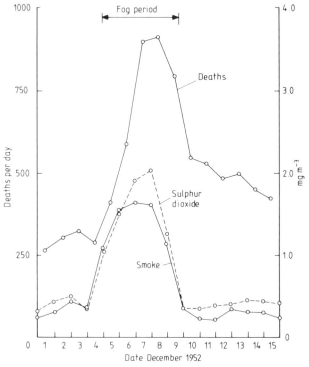

Figure 1.3 Daily air pollution and deaths in the London smog, December 1952. From Wilkins (1954)

†The word smog (smoke + fog) is usually attributed to Dr Des Voeux around 1910.

The most immediate result of the 1952 London episode was the establishment of the Beaver Committee. Their recommendations led directly to the Clean Air Act of 1956—a landmark in British legislation. This empowered local authorities to establish smokeless (Clean Air) zones, thus attempting to eliminate smoke pollution from both industrial and domestic sources. The effect of the Clean Air Act has been augmented or, arguably, superceded, by the socioeconomic pressures resulting in the coincident trend away from the use of coal in domestic heating.

Thus both air pollution and the degree of perception of pollution as a problem have changed and will continue to change. Steam locomotives (figure 1.4) have been replaced by electric and diesel locos in many countries; oil has replaced coal in many industries and yet with the impending

Figure 1.4 Steam locomotives in the UK in the 1960s (top) and South Africa in the late 1970s (bottom).

shortage of oil a new chapter in fuel usage is beginning to be written. (The debate continues as to the relative merits of nuclear power, alternative or renewable resources and a return to a coal economy†—possibly including the reintroduction of domestic coal fires and steam locomotives.) Pollution from vehicles (hydrocarbons, carbon monoxide and oxides of nitrogen) is of increasing concern; as is the formation of photochemical smogs (Chapter 7) in urban areas. Increasing trends over the last few centuries, reported by many as being exponential, may have been severely moderated by the 'Oil Crisis' of 1972–3 and the impending 'Energy Crisis' of the 1990s. An increasing per capita energy use has been associated with a higher standard of living (and often a higher degree of energy wastage) and a developing industry and economy. An increasing total consumption is thus further related to the population size. The global population is still increasing, despite an almost zero (or even negative) growth rate in many northern hemisphere developed countries. Extrapolations are dangerous. Predicting future population on past trends (again observed to be almost exponential), taking into account the possible increase in family planning and different social attitudes towards smaller families, is difficult. Hodges (1973) presents the calculation by taking the reciprocal of population, p, plotted against time. When $1/p$ becomes zero, we have on this planet an infinite population. This, Hodges calculates, will occur on Friday, November 13 2026, when most of us will still be alive!

NOMENCLATURE

The *phase state* of an element or compound is often related to its weight; the larger and heavier a molecule is, the more likely it is to exist in the solid or liquid state (although this is by *no* means a hard and fast rule). The basic difference between the three basic states (solid, liquid and gas) is that of the velocity with which the constituent molecules (or atoms) are moving.

Gases
The *kinetic theory* describes a gas as widely separated molecules, moving at high speeds (e.g. atmospheric nitrogen has a mean velocity of about 500 m s^{-1}) in which there are frequent collisions between molecules. The gas mass has no shape (unless confined within a container) and little order. More ordering results from lower velocities and is exemplified in the liquid or, eventually, the solid state. (This decrease in velocity can be easily brought about by, for instance, decreasing the temperature.)

For a gas, one mole occupies a volume of $22.4 \times 10^{-3} \text{ m}^3$ ($= 22.4$ litres)

---

†The *World Energy Outlook*, recently published by the OECD, forecasts a doubling of coal production and use by the year 2000.

under standard conditions of temperature and pressure (STP) which are 273 K (0 °C) and 1.013×10^5 N m^{-2} (760 mm Hg). The volume at non-standard conditions can be calculated from the *ideal gas law*. This law combines Boyle's, Charles's and Gay-Lussac's laws to give

$$pV = RT \qquad \text{or} \qquad p = \varrho RT \qquad (1.1)$$

where p is the pressure, V the volume per unit mass, T the temperature, ϱ the density $(= 1/V)$ and R the gas constant. The value of R depends on the gas in question. However it can be replaced by the universal gas constant $R^* = mR$ (where m is the molecular weight of the gas under consideration)†. The value of this constant is 8314 J mol^{-1} K^{-1}. (For air the value of R is 287 J kg^{-1} K^{-1}.) The ideal gas law applies to and therefore defines an *ideal gas*. Unfortunately no gases are ideal; although the ideal gas law is a common and useful approximation.

Liquids

In a liquid, molecular velocities are low enough for the element or compound to have more 'shape' but still retain sufficient 'fluidity' to change shape in accord with the shape of the container. The rate at which a liquid flows is measured in terms of its *viscosity* or 'measure of stickiness'. Liquids with high viscosity, such as treacle, flow slowly. Quantitatively, viscosity, represented by the Greek letter eta (η), is measured in m^2 s^{-1} or centistokes (1 centistoke = 0.01 m^2 s^{-1}); although in measurement of oil viscosities (see Chapter 6)—important in determining combustion characteristics—the obsolescent Redwood scale is still in use in some countries, including the UK.

Of great importance to air pollution meteorologists is the ability of liquid droplets to affect the direction of a light beam passing through them. Their physical presence in the atmosphere can result in light *scattering*, whereby quanta of light impinge upon the drop and are reflected away at some angle. The extent of scattering depends upon the size of the drop relative to the wavelength of the incident radiation, the droplet concentration and the physical characteristics of the scatterer. This effect is observable in terms of the *visibility* or the *range*—the distance to which one can see clearly (see Chapter 5). Visibility decrease was probably the most obvious indicator of the presence of air pollutants to the man-in-the-street in the days before clean air legislation and the more recent concern for the environment.

Two other features of liquids which are of some importance here are the vapour pressure and surface tension. Some higher velocity molecules in a liquid will tend to escape to form a gas or *vapour* above the liquid. (The terms gas and vapour have no significance chemically. Gas is usually applied to a chemical found normally in this phase; a vapour is the gas

---

†For dry air, $m = 28.97$.

phase of a chemical which exists as a liquid at STP.) This vapour exerts a pressure on the liquid surface known as the *vapour pressure*. Such considerations are important in determining the rate of growth and decay of raindrops which are implicated in the removal of some atmospheric pollutants (see Chapter 3). Also of some interest in cloud and aerosol formation is the *surface tension*. This is the resultant force arising from an imbalance of molecular attractive forces for the molecules near the surface of a liquid drop—there is a greater attraction radially inwards, contracting the surface to the smallest possible area. The (finite) radius of curvature afforded by a pollutant particle assists in providing a nucleus for condensation by reducing the surface tension necessary for the aerosol to form. The surface tension force is measured in $N m^{-1}$.

Solids
When a liquid freezes it solidifies. The molecular structure is well defined—the molecules are often arranged in a strict geometrical pattern, although some movement occurs. (Molecular motion only ceases at absolute zero.) Solids in the atmosphere also have scattering properties and may (as also may liquids) absorb light and other pollutants. Sorption properties of solids are often utilised in air pollution control devices (see Chapter 8). Particles may act as catalysts or provide an air pollution hazard in themselves. In this context they are more usually referred to as *particulates*.

Other terminology
Other terms are also used to describe air pollutants. These are often ill- or loosely defined. Several have either been taken from or given to common usage. It is useful, however, to attempt a definition for these terms although the list cannot promise to be comprehensive and there are shades of meaning and/or application in different countries.

Primary pollutants are those present in the atmosphere mainly as a result of direct emission e.g. particulates, sulphur dioxide (SO_2), carbon monoxide (CO).

Secondary pollutants are those present in the atmosphere mainly as a result of chemical reactions e.g. peroxyacetylnitrate (PAN), ozone (O_3), sulphur trioxide (SO_3).

Grit: solid particles greater than 76 μm (UK) or 100 μm (EEC). Particles of size greater than about 500 μm are seldom airborne.

Fly ash: particles of ash entrained in the waste gas from a combustion chamber.

Dust: solid particles (often generated by ore crushing and grinding) which settle under gravity—usually taken to be in the size range 1–76 μm.

Respirable particles: particulates which may be taken into the pulmonary system (assumed to be less than about 3–5 μm).

Smoke: carbonaceous particles (usually less than 1 μm) resulting from incomplete combustion of e.g. coal, tar, oil, tobacco.

Fume: solid particles generated by condensation (often after volatilisation from molten metals)—size range less than 1 μm.

Aerosol: a suspension of liquid or solid particles (usually less than about 50 μm in diameter).

Fog: an aerosol of liquid droplets at or near ground level (usually of natural origin).

Mist: suspension of liquid droplets generated by condensation or by dispersion from a liquid state.

Haze: aerosol which restricts visibility (less so than for a fog or (meteorological) mist which also has a much higher ($\sim 100\%$) relative humidity).

Smog: smoke + fog. Now also used in the term *photochemical smog* to describe the mixture of oxides of nitrogen, ozone and complex organics created by the action of sunlight on urban pollutants (notably in Los Angeles and Tokyo).

Gases: fluids in the gaseous phase at standard temperature and pressure (STP). Gases diffuse.

Vapours: fluids in the gaseous phase, but normally in the liquid or solid phase at STP. Can be liquefied by increasing pressure or decreasing temperature alone. Vapours also diffuse.

Concentrations

Concentrations may be measured in μg m^{-3} or mg m^{-3}. However, concentrations may be alternatively expressed in ppm (parts per million) or ppb (parts per billion (10^9)). This is really a smaller unit of the same type as % by volume ($^v/_v$) and gives a number ratio. It is most useful for gases (i.e. number of molecules per given number of molecules of air) but not for particulates, since it is not primarily the number, but the total mass that is the important parameter. Unfortunately the conversion (for gases) between ppm and μg m^{-3} is not straightforward, since it depends upon the temperature and pressure of the gas. The conversion is undertaken as follows.

At standard temperature and pressure (273 K and 1.013×10^5 N m^{-2}), M kilograms of gas of molecular weight M occupy 22.4 m^3. If however only p parts per million of this volume are occupied by this pollutant gas (i.e. a fraction of 10^{-6}) then we have $M \times 10^{-6} \times p$ kilograms of pollutant gas occupying 22.4 m^3 (the remainder of the gas is assumed to be air). Thus 1 m^3 contains

$$\frac{M \times 10^{-6} \times p}{22.4} \text{ kg}$$

$$= \frac{M \times p \times 10^3}{22.4} \mu\text{g}.$$

Hence

$$p \text{ ppm} \equiv 1000 \, Mp/22.4 \, \mu g \, m^{-3} \qquad \text{(at STP)}.$$

(Values at other temperatures are derived by using the ideal gas law, equation (1.1).) Some typical conversions are given in table 1.2.

Table 1.2 Equivalent values for concentrations of various species in $\mu g \, m^{-3}$ equivalent to 1 ppm $^v/_v$ for two different temperatures at a constant pressure of $1.013 \times 10^5 \, Nm^{-2}$.

	273 K	298 K
SO_2	2860	2620
CO	1250	1145
NO		1230
NO_2		1880
O_3	2141	1962
PAN [$CH_3(CO)O_2NO_2$]	5398	4945

CHEMICAL KINETICS

The two fundamental questions we need to answer in studying air pollution chemistry are whether there will be a chemical reaction and if so at what rate it will occur. The questions are answered by considering energy transformations in accord with the laws of thermodynamics, which say that energy can neither be created nor destroyed but simply changed and that the overall 'disorder' or *entropy* of a system must increase. These laws apply to *reactants* in all phases. A reaction involving atoms and/or molecules in the same phase is termed *homogeneous*. If two or more phases are involved the reaction is a *heterogeneous* one.

A reaction between a number of elements or compounds can be simply represented by a chemical equation which relates the appropriate ratios of species that will react together. For example, two molecules of carbon monoxide (CO) react with one molecule of oxygen (which contains two atoms) to give two molecules of carbon dioxide (CO_2). The equation which represents this reaction is written

$$2 \, CO + O_2 \longrightarrow 2 \, CO_2.$$

This is valid for any number of molecules such that the numbers are in the same ratios as represented by the equation i.e. a ratio of $2 : 1$ ($CO : O_2$).

The rate of a reaction is determined as the rate of change of concentration. At a constant temperature this rate is usually proportional to some power of the concentration of the reactants (A, B) such that

$$\text{rate} = k \, [A]^x [B]^y$$

where [ ] indicates concentration and k is the *rate constant* (per unit volume) and is a function of temperature. The exponents x and y are determined experimentally. The sum of these exponents is called the *order* of the reaction. The time for 50% depletion or *half life* is often used as a measure of the rate of a first-order reaction; the shorter the half life, the faster the reaction.

Table 1.3   Heat of formation of compounds at 298 K (25 °C)

	ΔH	
	kJ mol^{-1}	(kcal mol^{-1})
Gases		
HCl	− 92	(− 22)
H$_2$O	− 243	(− 58)
CO	− 112	(− 27)
CO$_2$	− 397	(− 94)
NO	+ 90	(+ 22)
NO$_2$	+ 33	(+ 8)
SO$_2$	− 298	(− 71)
H$_2$S	− 20	(− 5)
NH$_3$	− 47	(− 11)
CH$_4$	− 76	(− 18)
C$_2$H$_2$	+ 226	(+ 54)
CH$_3$OH	− 203	(− 48)
C$_2$H$_5$OH	− 239	(− 57)
Liquids		
H$_2$O	− 287	(− 68)
HNO$_3$	− 176	(− 42)
H$_2$SO$_4$	− 797	(− 190)
C$_6$H$_6$	+ 49	(+ 12)
CH$_3$OH	− 242	(− 58)
C$_2$H$_5$OH	− 282	(− 67)

Inherent in every molecule is an amount of energy: the energy required as input in order to form a molecule from its constituent atoms. This is the *heat of formation* (see table 1.3). For example, creation of acetylene from carbon and hydrogen requires 226.4 kJ (53.9 kcal) of energy for every mole. Many molecules, especially oxides, release heat during their formation (hence the negative signs in table 1.3)†. Often symbolised as ΔH, the heat of formation can give information not only on how much energy is released during a reaction, but also on what the more likely products will

---

†Some authors define ΔH as the heat *liberated* during a reaction in which case all the signs in table 1.3 would be reversed e.g. ΔH for HCl would be 92 kJ mol^{-1} (22 kcal mol^{-1}) heat *released*.

be. For instance, the heat of formation of ferric oxide, Fe_2O_3, is $-833.7\,kJmol^{-1}$ ($-198.5\,kcalmol^{-1}$), whilst that of ferrous oxide, FeO, is $-270.1\,kJmol^{-1}$ ($-64.3\,kcalmol^{-1}$). As both these values of ΔH are negative, it indicates that heat is liberated during the oxidation of iron. More heat is given out in the formation of Fe_2O_3, which implies that this oxide is more likely to be formed, due to the propensity of reaction products to attain the lowest possible energy state. A graph of such a reaction, if it occurred *spontaneously*, is given in figure 1.5, in which the energy levels are plotted as a function of time. The energy of the products of the reaction is, in total, less, and the amount of energy liberated is easily calculable as the (vertical) difference in energy levels. Such spontaneous reactions seldom occur, although if the ambient conditions are 'correct' they may. For example paper spontaneously combusts at a temperature of 506 K (451 °F). More usually there is an energy barrier (see figure 1.6) whereby an additional amount of energy (known as the *activation energy*) must be input to the system before any heat is released. For an activation energy E_a and a net heat release (as if spontaneous) of $-\Delta H$ (since $\Delta H < 0$ here), the energy released by the reaction is $E_a - \Delta H$; but as E_a was needed

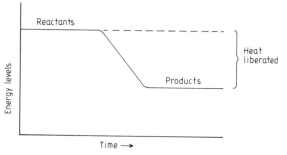

Figure 1.5 Energy levels during (spontaneous) chemical reaction.

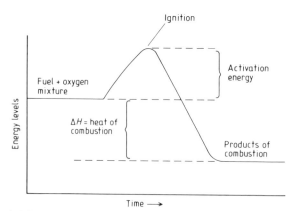

Figure 1.6 Energy levels for chemical reactions such as combustion.

to start the reaction, only $-\Delta H$ is available for use (see Chapter 6). One of the modes of operation of a *catalyst* may be in reducing the required activation energy. The input energy may be in the form of radiation; giving rise to a *photochemical reaction*. The amount of energy available depends on the wavelength of the radiation but is only available to the molecule *after* absorption. Absorption of a single quantum of radiation gives an energy E of

$$E = 12.008 \times 10^{-5}/\lambda \quad \text{kJ mol}^{-1} \tag{1.2}$$

where λ is in metres.

ATMOSPHERIC REACTIONS

Atmospheric chemistry parallels the reactions of classical chemistry, but is a mélange of many reaction possibilities—hence the interest and challenge. Some of the 'peculiarities' are as follows.

(*a*) The concentrations of reactants may be extremely low (less than $100 \ \mu\text{g m}^{-3}$ in many instances).

(*b*) Concentrations change as a result of variability of emissions.

(*c*) Meteorological effects can alter concentrations over many time and space scales usually inhomogeneously.

(*d*) The phase may depend upon the concentration, pressure and temperature.

(*e*) The type of pollutant may vary in accord with emissions and atmospheric reactions.

(*f*) Pollutant removal is effected and affected by several chemical and meteorological factors.

SOURCE TYPES

A matrix approach is the only method really suited to the study of air pollutants; yet by its very nature a book must be linear. Pollutants may be classified either by type of emission (e.g. industry, domestic) or by chemical species (e.g. SO_2, NO_2) and then the cross reference made. Some duplication is often necessary. In this book we will discuss the chemistry of specific pollutants later in this chapter without including a long list of industries. Some attempts at synthesis will be made briefly in the last section of this chapter under the heading Emission Inventories and in later chapters.

For the present it is sufficient to observe that over 90 % of primary pollutants originate from combustion processes (see Chapter 6). Emitters can be roughly divided into *stationary sources* (e.g. industry, commerce, domestic) and *mobile sources* (notably transport) (Chapters 6 and 7 respectively). The two types not only have different associated chemical species, but have a different nature—the former being identifiable spatially and the latter a much more varying phenomenon.

POLLUTANT TYPES

The relative 'importance' of each of the following pollutants in terms of environmental health, aesthetic effects and social attitudes will be discussed in later chapters. Initial concern was directed towards smoke and unburned hydrocarbons; at the time of writing there is concern, in various countries, about sulphur dioxide, lead particulates and photochemical smogs. It is impossible to predict future social attitudes, which may or may not reflect changing scientific opinion. For example, in Britain, Environmental Health departments receive an increasing number of complaints about odours—yet seldom are these as damaging to health as many pollutants (e.g. lead, SO_2) which often pass unnoticed. The following order is thus arbitrary in discussion of the chemistry of many of the compounds involved in air pollution.

Sulphur oxides (SO_2 and SO_3)

Of the two oxides of sulphur (sulphur dioxide and sulphur trioxide) sulphur dioxide is by far the more important as a primary pollutant. There is a natural background SO_2 concentration of around 0.001 ppm (a global average—see table 1.1). This is mostly of volcanic origin or results from oxidation of hydrogen sulphide released as a result of anaerobic decomposition

$$2\,H_2S + 3\,O_2 \longrightarrow 2\,SO_2 + 2\,H_2O.$$

Anthropogenic production of SO_2 centres on the combustion process since nearly all fossil fuels contain sulphur as an impurity—see Chapter 6 for details. Much attention is focused today on sulphur dioxide as an indicator of pollution, largely because it is emitted in sufficiently large quantities (see table 1.4) to be easily detectable. In addition, measuring techniques for quantification are relatively well developed. Local concentrations can be high, despite the low retention or residence† time (see table 1.5). If emitted high enough, large distances can be travelled before removal and the problem becomes one of international concern.

Chemically, sulphur dioxide is a non-inflammable colourless gas. However it can be detected both by taste and smell at concentrations as low as around $1000\ \mu g\,m^{-3}$. (This detection threshold varies markedly from person to person—see Chapter 5.) It is extremely soluble in water; at room

---

†The residence time, τ, is the time taken for 100 % depletion assuming continuous replacement. It can be related to the half life, $\tau_{1/2}$, by $\tau = \tau_{1/2}/\ln 2$ for a first-order reaction. The half life $\tau_{1/2}$ is defined as the length of time for 50 % depletion without replenishment.

temperature (about 293 K or 20 °C) 1 kg of water will take into solution over 0.1 kg of SO_2.

Table 1.4 Worldwide emissions (annual estimates) of primary pollutants. Information compiled from various data sources.

| Pollutant | Anthropogenic | | Natural | |
	Major source	Emission (10^6 tonnes)	Major source	Emission (10^6 tonnes)
SO_2	Fossil fuel combustion	196	Volcanoes	O(5)
CO	Incomplete combustion (including vehicles)	275	Forest fires	75
CO_2	Combustion	1.4×10^4	Biological decay, release from oceans	10^6
NO	Combustion	53	Bacterial action in soil	430
NO_2	Combustion			658
N_2O	None	0	Biological action in soil	590
NH_3	Waste treatment	4	Biological decay	1160
H_2S	Sewage treatment	3	Volcanoes, anaerobic decomposition	100
HC	Combustion, chemical processes	88	Biological processes	480
Particulates	Fossil fuel combustion, slash and burn agriculture	690	Sea-salt spray, volcanoes	1150–1500

In the atmosphere SO_2 reacts with more oxygen to form the trioxide (which is thus a secondary pollutant)

$$SO_2 + O + M \longrightarrow SO_3 + M$$

where M is a third species (such as NO_2, metallic oxide) which acts as a catalyst. This reaction is slow (several days) but can be speeded up by moisture or by impurities which can have a catalytic effect. Sulphur trioxide has a strong affinity for water, condensing as droplets of sulphuric acid

$$SO_3 + H_2O \longrightarrow H_2SO_4$$

although the chemical mechanism is as yet imprecisely understood (Levine and Allario 1982). An alternative path may be the reaction (of SO_2) with hydroxyl radicals or with naturally occurring hydrogen peroxide

$$SO_2 + H_2O_2 \longrightarrow H_2SO_4.$$

Table 1.5 Half lives of some gaseous pollutants. The residence time, for a first-order reaction, is equal to $1.44 \times$ half life.

Pollutant	Half life
SO_2	4 days
H_2S	2 days
CO	<3 years
NO/NO_2	5 days
NH_3	7 days
N_2O	4 years
HC	16 years (CH_4)
CO_2	2–4 years
Lead aerosol (inorganic)	7–30 days

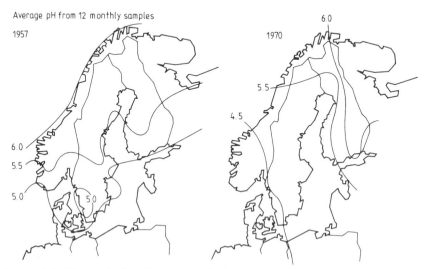

Average pH from 12 monthly samples

1957

1970

Figure 1.7 The trend of acidity in precipitation over Scandanavia over two decades. Redrawn from Likens (1976) by permission of the American Chemical Society.

These droplets are then similar to cloud water droplets. Hence the rain may contain not only water drops but also droplets of sulphuric acid. So-called 'acid rain' (containing not only sulphuric acid but also nitric acid derived from oxides of nitrogen) is accused locally of eating away car bodywork and clothes and internationally of lowering the pH of rivers and lakes (especially in Scandinavia—see figure 1.7), drastically altering the aquatic environment. There is at present much concern about the tall stack emissions from, particularly, the UK, the Ruhr and eastern Europe (affecting Scandinavia) and the USA (affecting Canada). The high level plume from the INCO nickel smelter at Sudbury, Ontario (currently, at 381 m, the highest chimney in the world (figure 1.8) and until very recently emitting up to 3 % of the world's sulphur) has been tracked as far away as Florida. Local deposition (within a 50 km radius of the chimney) is of the order of 10–20 %, although up to 100 % of pollutants may be removed during rainstorms (Chan *et al* 1982).

Figure 1.8 The chimneys at Sudbury, Ontario. The 381 m chimney (centre) is seen soon after construction i.e. before operation.

Other sulphur compounds

Hydrogen sulphide, emitted naturally, is toxic at concentrations of 0.1 % over a short period. It is, however, rapidly removed by oxidation to sulphur dioxide (a reaction also used industrially, see Chapter 8).

Mercaptans are a group of organic sulphur compounds with low odour thresholds. Ethyl mercaptan (C_2H_5SH) is detectable at concentrations as low as 1 part in 5×10^{10}. Mercaptans are produced at sewage works, food processing plants, by brick burning and wood pulp mills.

Oxides of nitrogen (NO$_x$)

There are many known nitrogen oxides, including nitrous oxide, N_2O or 'laughing gas'. Only nitric oxide (or nitrogen monoxide) NO and nitrogen dioxide NO_2 are emitted anthropogenically in large enough quantities to be of interest here. Both are formed by combustion at high temperatures. The nitrogen may be present as a trace impurity in fuel or in the air being used in the combustion process. The reactions can be summarised as

$$N_2 + xO_2 \rightleftharpoons 2 NO_x.$$

As can be seen in this reversible reaction, the specific oxide formed depends on the oxygen available. It also depends on temperature. Atmospheric conversion of NO to NO_2 does take place and is an important step in the formation of a photochemical smog (see Chapter 7). Typical peak, urban concentrations may be of the order of 1 ppm of NO and less than 0.5 ppm NO_2.

NO_2 is also implicated in respiratory disease in children as a result of emissions from gas cookers. This indoor air pollution may be growing as a result of the lack of ventilation in modern day dwellings, consequent upon the demise of the coal fire.

Chemically, nitric oxide (NO) is a colourless gas for which blood haemoglobin has a strong affinity. Nitrogen dioxide, on the other hand, is a reddish gas detectable (by its odour) at concentrations of about 0.12 ppm. It is a strong absorber of ultraviolet radiation.

Other forms of nitrogen

The other major nitrogen compound is ammonia (NH_3). This is a colourless, highly soluble, pungent gas with both natural and industrial sources (see table 1.4) and a residence time of about 7 days (table 1.5). In the atmosphere it forms salts, such as ammonium sulphate and ammonium nitrate, e.g.

$$H_2SO_4 + 2 NH_3 \longrightarrow (NH_4)_2SO_4.$$

Locally there may be hydrolysis of the ammonium sulphate aerosol to sulphuric acid causing irritation to the lungs.

Nitrogen compounds are removed from the atmosphere by oxidation described by the set of equations

$$NO + O_3 \longrightarrow NO_2 + O_2$$
$$NO_2 + OH + M \longrightarrow HNO_3 + M$$

$$\text{or} \quad \begin{cases} NO_2 + O_3 \longrightarrow NO_3 + O_2 \\ NO_3 + NO_2 \rightleftharpoons N_2O_5 \\ H_2O + N_2O_5 \longrightarrow 2 HNO_3 \end{cases}$$

$$NH_3 + HNO_3 \longrightarrow NH_4NO_3$$

(see also the discussion on NO_x/O_3 chemistry in Chapter 7). Some HNO_3 is also removed directly by rainout and washout (acid rain) or by dry deposition. Its role in acid rain production is currently being evaluated. Current estimates (Levine 1984) suggest that about one third of the acidity in rainfall is caused by the presence of nitric acid (the remaining two thirds by sulphuric acid).

Oxides of carbon (CO and CO_2)

The study of carbon oxides is an interesting one for several reasons. Carbon monoxide is poisonous—it has a high affinity for haemoglobin (Hb). The reaction

$$Hb + CO \longrightarrow COHb$$

leads to the formation of carboxyhaemoglobin (COHb) in the blood. This reaction prevents the normal formation of oxyhaemoglobin (O_2Hb)

$$Hb + O_2 \longrightarrow O_2Hb$$

in which oxygen is transferred, in the lungs, to the bloodstream and then delivered to the requisite cell site by the reverse reaction

$$O_2Hb \longrightarrow Hb + O_2.$$

Cells deprived of oxygen function poorly or may die since COHb in the blood cannot be removed, except when the haemoglobin is replaced *in toto* by natural processes (over a period of several days). Concentrations of CO of 250 ppm begin this process and the victim loses consciousness if exposed to such a concentration for more than a few hours. If the concentration rises to around 1000 ppm, death is quick. At lower concentrations, such as those found in urban areas (see figure 1.9), driving judgement may be impaired. However a traffic survey in London showed that a greater influence on blood COHb levels was whether the interviewee was a smoker or a non-smoker (table 1.6).

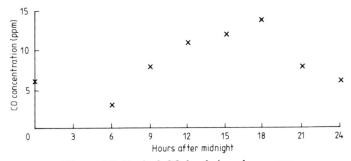

Figure 1.9 Typical CO levels in urban areas.

Table 1.6 CO levels in blood of London taxi drivers. ©The Open University Press from PT 272 units 13–14.

	Day drivers		Night drivers	
	Sample size	% COHb mean (range)	Sample size	% COHb mean (range)
Non-smokers	8	2.3(1.4–3.0)	12	1.0(0.4–1.8)
Smokers	19	5.8(2.0–9.7)	10	4.4(1.0–8.7)

Carbon monoxide is a product of *incomplete combustion*—that is combustion in which carbon compounds are only partially oxidised (i.e. to CO) and could be further oxidised to CO_2, which is therefore a product of *complete combustion*. (No further oxidation is possible as CO_3 does not exist.) Since carbon monoxide is toxic, whilst carbon dioxide is not, industry always aims to achieve complete combustion. Indeed, throughout most of this book we regard CO as 'bad', CO_2 as 'good'—at least CO_2 is not *as* bad, because its effects may not be apparent for decades. During the last twenty years some concern has arisen about increasing *global* levels of carbon dioxide (figure 1.10). Since the Industrial Revolution CO_2 levels seem to have been increasing at a rate of about 0.7 ppm per year. The problem arises in the 'greenhouse effect' by which the 'blanket' of CO_2 warms the Earth (see Chapter 2 for full details).

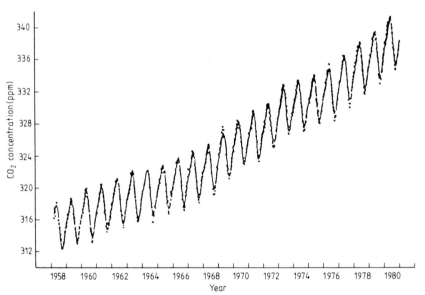

Figure 1.10 Atmospheric CO_2 concentrations (weekly average) measured at Mauna Loa, Hawaii. The curve is a spline fit assuming an annual variation of constant amplitude superposed on a long-term trend. Redrawn from Rycroft (1982) by permission.

Both CO and CO_2 are natural components of the atmosphere (see table 1.1). Chemically they are formed by direct oxidation

$$2\,C + O_2 \longrightarrow 2\,CO$$
$$2\,CO + O_2 \longrightarrow 2\,CO_2.$$

This latter reaction is relatively slow, needing large amounts of input energy. Both oxides are colourless gases and reasonably inert. Carbon dioxide does not support combustion and is thus often used in fire extinguishers.

Ozone

Ozone (O_3) is a gas which occurs naturally in the stratosphere (see Chapter 2), but can act as an irritant at anomalously high concentrations at ground level, where it may be formed in the NO_x/O_3 photochemical smog cycle (Chapter 7) by the reaction between O and O_2. At or near ground level ozone concentrations are usually associated with such photolytic reaction series (the O is a result of photolytic dissociation of NO_2) and hence ozone is typical of areas of high irradiation (together with a high degree of urbanisation). Until relatively recently high ozone concentrations were confined to areas such as Los Angeles. However, decreasing atmospheric aerosol loadings have permitted other urban areas to receive an increased amount of solar radiation. In 1976 ozone concentrations measured at five sites in southeast England (Apling *et al* 1977) were observed well in excess of the United States Environmental Protection Agency (USEPA) standard and Greater London Council guidelines of 80 ppb. Table 1.7 shows that in this year the highest mean hourly concentrations were in excess of 200 ppb. The high concentration increasingly observed in the rural environment is attributed to advection of ozone from industrialised northwest Europe.

Table 1.7 Highest hourly mean ozone concentrations (ppb) observed during 1972–6 (data from Apling *et al* 1977). Figures refer to ozone concentrations in ppb and are followed by the original reference.

Year	Central London	London suburbs	Rural Southern England
1971	—	—	101 Atkins *et al* (1972)
1972	112 Derwent and Stewart (1973)	—	128 Bell and Cox (1975)
1973	136 Cox *et al* (1975)	—	141 Cox *et al* (1975)
1974	164 Derwent *et al* (1976)	—	120 Cox *et al* (1976)
1975	150 Ball (1976)	130 Ball (1976)	177 Cox *et al* (1976)
1976	210 Apling *et al* (1977)	210 Apling *et al* (1977)	258 Apling *et al* (1977)

Hydrocarbons (HC)

Hydrocarbons is the general name for a chemical containing both hydrogen and carbon. They may be open chain or cyclic. *Open chain* hydrocarbons are usually single chains, but may contain one or more double bonds. These *unsaturated* hydrocarbons can be identified as belonging to groups or families, in which different members have different numbers of carbon atoms in the 'backbone'. For example, the simplest *alkene* (olefin)—having the general form C_nH_{2n}—is ethene (ethylene)

$$\begin{array}{ccc} H & & H \\ \diagdown & & \diagup \\ & C=C & \\ \diagup & & \diagdown \\ H & & H \end{array}$$

The double bond (although strong) may be broken by further hydrogenation. When all double bonds are broken the hydrocarbon is said to be *saturated*. Two of the simpler *alkanes* (paraffins)—with the general form C_nH_{2n+2}—are methane (CH_4) and propane (C_3H_8)

$$\begin{array}{cc} \begin{array}{c} H \\ | \\ H-C-H \\ | \\ H \end{array} & \begin{array}{c} H \quad H \quad H \\ | \quad\; | \quad\; | \\ H-C-C-C-H \\ | \quad\; | \quad\; | \\ H \quad H \quad H \end{array} \\ \text{methane} & \text{propane} \end{array}$$

If the carbon chain forms a closed loop of six carbon atoms, the structure is known as the benzene ring, symbolised ⬡ in which the bonding electrons cannot be uniquely identified with any specific carbon-to-carbon bond (delocalisation of electrons). Again *cyclic hydrocarbons* may be saturated or unsaturated. Benzene itself (C_6H_6) is unsaturated; whilst hydrogenation to C_6H_{12} saturates the molecule.

Hydrocarbons are of specific interest in air pollution studies as a result of the worldwide use of hydrocarbon fuels in the internal combustion engine. Petrol (gasoline) is a complex mixture of various hydrocarbon fractions. Molecular weights therefore vary as does the volatility; although simple observation at a petrol station will indicate the relatively easy and rapid volatilisation. This gives rise to the strong possibility of loss to the atmosphere from anywhere that petrol is stored in containers (including petrol tanks in cars) which must be vented to avoid the buildup of vapour pressure. Only hydrocarbons containing more than about twelve carbon atoms (i.e. molecular weights in excess of approximately 150) are not subject to this rapid vapourisation. It is also possible for hydrocarbons to leave the combustion chamber unburnt, especially if the fuel to air ratio is too high.

An important group of hydrocarbons is the aromatic hydrocarbons or arenes, with the general formula C_nH_{2n-6} (where $n \geqslant 6$). The simplest are benzene (C_6H_6) and methylbenzene or toluene ($C_6H_5CH_3$). Polycyclic aromatic hydrocarbons (PAH) are currently of great concern (Daisey and Lioy 1981) arising from vehicular combustion and oil burning for space heating. These include known carcinogens, notably benzo(a)pyrene.

Hydrocarbons are implicated in the series of chemical reactions with oxides of nitrogen, catalysed by sunlight, that produce a photochemical smog. Some of the resultant pollutants are complex organic molecules, for example, acrolein H_2C_3OH

$$\begin{array}{c} H \qquad\quad |H \\ \diagdown C = C \diagup \\ H \diagup \qquad | \\ \qquad\quad C = O \\ \qquad\quad | \\ \qquad\quad H \end{array}$$

acrolein

and PAN (peroxyacetylnitrate)

$$\begin{array}{c} H \quad O \\ | \quad\; || \\ H - C - C - O - O - NO_2 \\ | \\ H \end{array}$$

(see also Chapter 7).

Particulates

In air pollution, the term particulates is used to describe anything larger than a single molecule (i.e. size greater than about 2×10^{-10} m or 0.0002 μm) in either the liquid or the solid phase. These may be produced naturally (e.g. hail, snow, fog) or be of anthropogenic origin (e.g. smoke, consisting of a suspension of fine carbonaceous particles and minute droplets of tar). Particulates are often subdivided according to size range. Particles of grit are those between 76 and about 500 μm which sediment rapidly. Dust also settles under gravity; particles being in the size range 1–76 μm. Smaller particles ($\leq 1\,\mu$m) tend to remain suspended for sufficient periods to form identifiable aerosols. The size distribution of atmospheric aerosols is generally regarded as being trimodal (figure 1.11). Particle growth mechanisms, such as condensation of gases together with larger particles, are evident.

The removal rate of particulates depends upon the density and size of the particle according to the well known Stokes's law (see below). This applies to particles large enough to have a significant fall speed.

Particulates greater than about 5 μm (figure 1.10) are not generally regarded as pollutants due to this gravitational removal process. At the other end of the size spectrum, particulates smaller than about 0.7 μm can suffer random molecular interactions—bombardment by air molecules resulting in collisions—observed as *Brownian motion* which tends to delay any gravitational action. For such smaller particles removal from the atmosphere is also effected by collision, washout and rainout. Figure 1.12 illustrates sizes typical of other small objects to put the above discussion into a wider context.

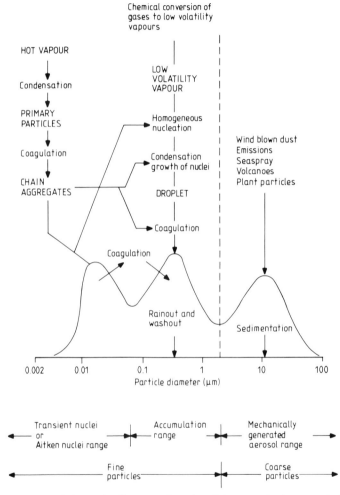

Figure 1.11 Schematic diagram of a trimodal atmospheric aerosol size distribution showing the principal modes, main sources of mass for each mode and the principal processes involved in inserting mass and removing mass from each mode. Redrawn from Whitby (1977).

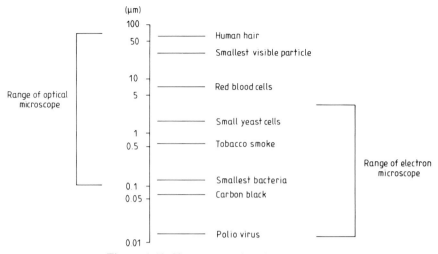

Figure 1.12 Size ranges of various particles.

Stokes's law is derived as follows. Consider a sphere of radius r and density ϱ falling through a medium of density ϱ_0. There are two forces acting on the sphere: the net (negative buoyancy) force, F_1, which is the volume multiplied by the difference in densities multiplied by the acceleration due to gravity, g, and the force due to the viscosity of the air, F_2 (a frictional effect), given by

$$F_2 = 6\pi\eta r V \tag{1.3}$$

where V is the velocity. Initially $F_1 > F_2$; so the particle accelerates until $F_1 = F_2$ after which the forces balance and a final steady speed V_0 (the *terminal velocity*) is reached.

$$V_0 = 2\ g\varrho r^2 / 9\ \eta \tag{1.4}$$

where η is the coefficient of viscosity (of air).

For small particles ($\lesssim 30\,\mu$m) the Cunningham correction, C, must premultiply V_0 to allow for the ability of such sized particles to 'slip through' the air molecules.

$$C = 1 + \{\lambda[1.26 + 0.4 \exp(-1.1\ r/\lambda)]/r\} \tag{1.5}$$

where λ is the mean free path of the molecules which is thus a function of altitude i.e. pressure (at STP λ is approximately 10^{-7} m). For a particle of size 0.1 μm, C is approximately 1.9, i.e. the terminal velocity is approximately double that calculated by the uncorrected formula (equation 1.4).

Furthermore, if the acceleration phase takes a short time in comparison with the overall settling time, it may be assumed that the particle has fallen with a constant velocity V_0 and hence the time, t_h, to fall a height h is calculated using the approximate equation

$$t_h = h/V_0 \qquad (1.6)$$

which slightly underestimates the settling time.

Calculations of such deposition rates (important in their own right) must take into account the random effects of turbulent motions. This is well illustrated in table 1.8 where it can be seen that, under the specified meteorological conditions, particles of 2 μm settle out slowly (only 5 % at a distance of 2 km downwind of the source) whereas at the same distance almost 50 % of particles of 10 μm size have been removed from the atmosphere. Here the dust cloud is assumed to be at a given height with negligible thickness vertically. Under neutral conditions dust clouds are likely to be dispersed to a greater extent in the vertical than under stable conditions. At large distances from the source, the pollution will be mixed relatively homogeneously through the mixed layer (boundary layer). Hence under real conditions other parameters need to be taken into consideration in calculating deposition rates. The damage caused also depends in a complex manner on size (see Chapter 5, where deposition in the lung and the possible resultant damage are related strongly to particulate size). One simple effect is in the capacity of particles to cover surfaces. Suppose dust settles over an area in a layer one particle deep. Since larger particles have a high volume/cross-sectional area ratio they have a lower surface cover per unit mass. This is illustrated in table 1.9.

Table 1.8 Typical percentage of dust deposition (windspeed about 5 ms^{-1}).

Particle size (μm)	Distance downwind (m)			
	100	200	1000	2000
2	Negligible	0.5	2	5
5	1	2	8	16
10	3	7	28	49

Table 1.9 Covering ability of dust. Approximate surface area covered by 1 kg of particles of stated size (layer one particle deep).

Size (μm)	0.1	0.5	1.0	2.0	5.0
Surface area (m^2 kg^{-1})	24 000	8000	2500	1200	800

Measurement of deposition rates is difficult (see Chapter 4), but a simple experiment credited to Carey will demonstrate this. A sheet of white paper is placed on the ground, either indoors or outdoors, and covered with a set of child's building bricks or similarly sized objects (figure 1.13). These are removed one by one each day (i.e. one brick is removed every 24 hours).

The resulting pattern depicts the extent of particulate deposition. If *m* bricks are used over an experimental period of *m* days, then the *m*th brick position will suffer one day's pollution, whereas the first brick position collects pollution from all *m* days (see table 1.10).

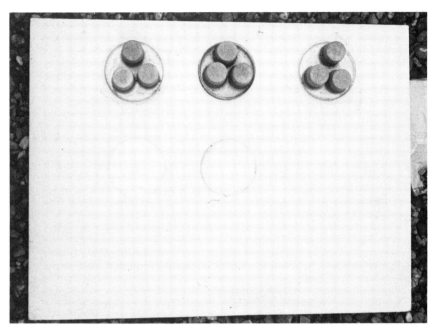

Figure 1.13 'Carey's experiment' using weighted Petri dishes, in which one dish is removed each day from a white surface to illustrate particulate pollution falling out of the atmosphere.

Table 1.10 'Carey's experiment' illustrating the relationship between brick removal and cumulative pollution.

Brick number	1,	2,	3, ...	*m*	experiment terminated at beginning of day $m+1$
Day of removal (beginning of day *i*)	1,	2,	3, ...	*m*	
Days of pollutant deposition		*m, m* − 1, *m* − 2, ...3, 2, 1			

Whilst in suspension in the atmosphere, particles can scatter and absorb light, reducing visibility and affecting both incoming solar radiation reaching the ground and the outgoing longwave energy balance.

Halogens

The halogens are the seventh group in the Periodic Table (see figure A.1 in the Appendix). Halogens are very reactive and hence can be troublesome (and dangerous to health) as air pollutants; although effects are localised in all but one case. *Fluorine compounds* are emitted from various industrial sources, notably brickworks. *Chlorides* (especially hydrogen chloride gas) may arise from, for instance, burning of PVC in incinerators. In some locations ammonium chloride emissions have caused problems. The only non-local pollutant is the set of compounds known as CFC (chlorofluorocarbons) or sometimes chlorohalocarbons. These have been used for some time as propellants in aerosol cans and have had widespread usage. In 1974 concern began to be expressed about these inert compounds when it was found that a reaction occurred between the photolysed CFCs and ozone and that, with their high residence times, it was feasible for them to enter the stratosphere and deplete the ozone layer. The discussion still continues with as many arguments for the immediate banning of CFCs as there are against. The USEPA recommended a discontinuation of their use in the USA and Canada. Sweden and a few other countries have done likewise. Scientists in the UK consider a ban premature and/or unnecessary. In any case stratospheric ozone levels are at present continuing to rise (see Chapter 2 for further details).

Lead

Atmospheric lead can be in one of two forms: organic or inorganic. Organic lead originates mainly from petrol (to which it is added as an 'anti-knock' agent in the form of tetraethyl or tetramethyl lead). Maximum permitted amounts are subject to changing legislation. The UK limit was set at 0.40 g dm^{-3} (grams per litre). Stricter controls were announced in May 1981 by which this level would be reduced to 0.15 g dm^{-3} by 1 January 1986 and agreement has been reached (1983), in principle, to eliminate lead from petrol by 1990; although EEC discussions on applying this to Europe are still (June 1983) not complete.

In West Germany the limit is set at 0.15 g dm^{-3} and in India approximately 0.5 g dm^{-3} of tetraethyl lead is added to petrol. In the USA, recently built cars use only lead free petrol compared to an added 0.343 g dm^{-3} in the mid-1970s.

In the internal combustion engine, the tetraalkyl leads are largely changed to inorganic forms by reaction with HCl and HBr. These latter compounds arise from combustion of 1,2-dichloroethane (ClCH$_2$CH$_2$Cl) and 1,2-dibromoethane (BrCH$_2$CH$_2$Br) which are also present in the petrol as *scavengers*. Consequently one of the major emissions is lead bromochloride (PbBrCl). These forms make up 70–75 % of the lead emitted from the exhaust, together with about 1 % of organic lead which has passed through the engine unchanged. The remainder of the lead is trapped within the exhaust system and by the engine oil. Organic lead emissions usually occur as vapours and the inorganic emissions as particles, often less than 1 μm in size (and hence respirable). Annual lead emissions from vehicles are typically 9000 tonnes (UK) and 130 000 tonnes (USA).

Inorganic lead also originates from various lead-based industries as well as coal combustion. Coal has a typical lead content of 17 ppm (UK) and 0–79 ppm (USA). Total inorganic emissions (non-vehicular) are of the order of 300 tonnes per year (UK) and 14 000 tonnes per year (USA).

Inorganic lead is a cumulative poison. If inhaled or ingested† it is readily absorbed and stored in the body, 90 % of body lead being retained in the bones. The half life of lead in bone is thought to be between 600 and 3000 days (and possibly up to 27 years in dense bone).

The now seldom seen classical symptoms of lead poisoning are a blue line around the gums, severe debility, anaemia and weight loss. Blood lead levels greater than about 80 μg per 100 ml are likely to produce severe symptoms. Recent studies (Yule *et al* 1981) suggest that in children levels of approximately 7 μg per 100 ml may impair intelligence—a level already attained in a large proportion of the juvenile population.

The metabolism of organic lead is different from inorganic lead—here the functioning of the nervous system may be impaired—although organic lead is not classed as a cumulative poison, some storage may occur in the liver in the form of trialkyl lead (Grandjean and Nielsen 1979).

The eight-hour TLVs (see Chapter 5) for tetraethyl lead, tetramethyl lead and inorganic lead are 100 μg m^{-3}, 150 μg m^{-3} and 150 μg m^{-3} respectively.

Other pollutants

Many other heavy metals are emitted, especially as a result of metal smelting. They can cause considerable localised pollution of the environment. *Cadmium* levels are frequently monitored around smelters. *Nickel* is responsible for causing cancers and dermatitis, although this is more of a problem within the industrial environment. *Mercury* is highly toxic, its vapours inducing 'madness' (hence the Mad Hatter in *Alice in Wonderland* —milliners used mercury in their work). *Beryllium* and *molybdenum* are also toxic metals associated with metal alloy hardening and smelting and have been held responsible for farm animal deaths.

Of more local or more infrequent occurrence and concern are *asbestos* fibres. Asbestosis is a killer disease with a high incidence in workers in the asbestos industry. The use of blue asbestos is now banned in many countries for this reason. Of more concern are the lung cancers and the tumorous mesothelioma associated with asbestos exposure. *Nerve gas*, on trial, has

---

†One of the more important sectors of the population at greatest risk is children who can be both exposed to urban airborne lead and ingest lead via contaminated objects (e.g. covered by roadside dust). Such effects are exacerbated in cases of *pica* when the child habitually eats dirt and chews possibly contaminated objects. The lead gained in these ways may approximately double the child's lead intake.

killed, in one case, several thousand sheep. *Aero-allergens* are not considered fatal but can cause high discomfort (usually in the form of 'hay fever') to those who have a high susceptibility. Ragweed pollen and plane tree pollen are perhaps as important as grass pollens.

Although, in total, cigarette (and cigar and pipe) *smoking* ranks as a minor source of pollutants, its localised effects can be very great. Damage is caused not only to the smoker from inhaled smoke but also to non-smokers subjected to a smoke-filled atmosphere and cigarette side-smoke. In fact the latter has been shown to be, in many ways, as dangerous as inhaled smoke as it is unfiltered and contains many carcinogenic compounds as well as toxins such as nicotine. It has been estimated that side-smoke alone causes two cases of lung cancer each year per 100 000 non-smokers.

Cigarette smoke (about 5×10^{12} particles of respirable size—about 0.2 μm—are produced by one cigarette) is composed of about 3000 different chemical compounds of which about 1000 are carcinogens (e.g. benzopyrene), some of which are in the gaseous phase (e.g. nitrosamines). Lung cancer is the major induced cancer, although the pancreas, lower urinary tract and upper digestive tract may also be affected. Blood levels of carbon monoxide are higher for smokers than non-smokers (resulting from about 70 mg CO per cigarette) and medical questionnaires now frequently begin with the question 'Are you a regular smoker?'

In the USA smoking is now prohibited in many public places (including public transport). In this respect, the UK and Europe lag behind and too often restaurants and theatres permit smoking. In enclosed spaces such as these and perhaps more especially lifts and waiting rooms many people are of the opinion that the freedom to pollute should be denied.

Smells and odours increasingly offer cause for complaint but it is often considered that they are fairly harmless. However they cause an aesthetic damage to our environment—and are often the centre of liaison work between air pollution control workers and the general public.

EMISSION INVENTORIES

Emission sources may be divided into natural and anthropogenic. The latter class can be further divided into stationary and mobile. The amounts of each of the different pollutants mentioned above can either be measured, calculated or inferred on a variety of time and space scales. It is common to look at total yearly emissions over a whole country (see, for example, figure 1.16 in which SO_2 and smoke emissions for the UK are plotted as yearly totals over the period 1950–76). This is not necessarily making the same inference about air pollution trends of, say, SO_2 at a single site, due to the contrasting time and space scales (see the later discussion about *observed* pollutant concentrations in Chapter 4).

On a national scale, it may be possible to look at sources divided into the three groups above or even further subdivided. In specific case studies it is more important to restrict one's area of interest to an even smaller locality (although the influence of long-range transport of air pollutants is much less well understood).

It is impossible, within the scope of a book such as this, to present all the data gathered. However it is of interest to present and discuss some data sets in order that the overall magnitude of the problem can be appreciated and put into perspective.

Estimates have been made of global production of both natural and anthropogenic pollutants. Some of these values are presented in table 1.4, based largely on figures for 1969.

It can be seen that, with two notable exceptions (SO_2 and CO), anthropogenic sources are negligible in comparison with natural sources. Then why the concern? Simply because, firstly, anthropogenic sources are extremely localised (only a small percentage of the global surface is industrialised) and, secondly, it can be assumed that the 'natural' world is in a steady state. Anthropogenic additions may tilt the balance. Whether the balance is restored or upset depends to a large degree upon whether the system is transitive or intransitive (Lorenz 1975) i.e. whether it possesses only one or more than one stable state.

The largest anthropogenic sources, again associated with industry and transport, are sulphur dioxide and carbon monoxide—and hence the detailed consideration of these pollutants in any book on air pollution.

For each of the figures in table 1.4, it is possible to partition the emissions between source types. Categories for sources may be more detailed than the simple breakdown of stationary, mobile and natural given above. This type of description is more often used for national or regional data. UK emissions of smoke are shown in figure 1.14. Traditionally SO_2 emissions have been classified as coming from power stations, domestic sources or others (figure 1.15). Today, the height of the emitter is realised to be a more important and useful discriminant (figure 1.16). Following trends away from domestic coal combustion to oil and gas usage and regional power generation the anomalously high emissions in figures 1.14 and 1.15 (e.g. that for 1979) are a result of increased space heating required in colder winters. UK emissions of oxides of nitrogen, carbon monoxide and hydrocarbons are given in table 1.11.

Similar figures are available for the USA for carbon monoxide, sulphur oxides, nitrogen oxides, hydrocarbons, particulates etc. Estimates for particulate emissions for 1976 were presented by Evans and Cooper (1980). Open sources (wind erosion of soils, vehicles, quarrying etc) contributed over 580×10^6 tonnes of particles to the atmosphere. Anthropogenic sulphur emissions in the USA are of the order of 30×10^6 tonnes per annum.

In comparison, Canadian SO_2 emissions for 1974 were 5.9×10^6 tonnes (Voldner *et al* 1980) and for Europe, 8.7×10^6 tonnes (Semb 1978).

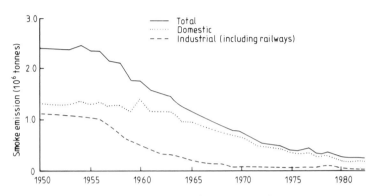

Figure 1.14 Trends in UK smoke emissions. Data from *Warren Spring Laboratory Reports.*

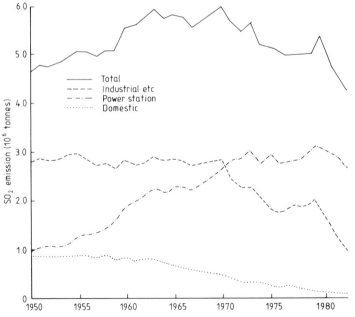

Figure 1.15 Trends in UK sulphur dioxide emissions. Data from *Warren Spring Laboratory Reports.*

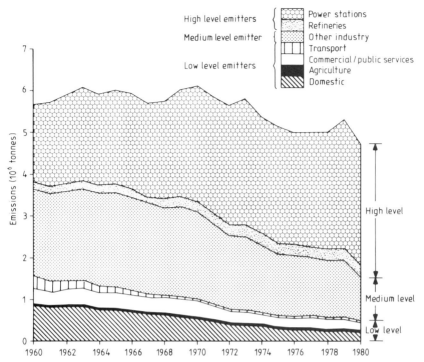

Figure 1.16 Trends in UK sulphur dioxide emissions as a function of height of emission. Data from *Warren Spring Laboratory Reports.*

Table 1.11   UK emissions of NO_x in 1000 tonnes NO_2 equivalent, CO in 1000 tonnes and hydrocarbons in 1000 tonnes for 1981. *Digest of Environmental Pollution and Water Statistics,* No 5 (1982).

Source	NO_x	CO	Hydrocarbons
Domestic	48	502	78
Commercial and industrial	401	54	26
Power stations	818	47	14
Industrial processes and solvent extraction			600
Incineration and agricultural burning	12	220	38
Road vehicles: petrol engined	309	7514	476
diesel engined	167	228	37
Railways	42	15	10
Totals	1797	8580	1279

The tabulated values presented are *emission inventories*—lists attributing values to specific source types. Any such list within a given and well specified region can be classified as such. They give gross estimates of the major pollutant types present in an area and can thus give a good indication as to what sort of monitoring network could be most usefully employed in any investigation of the air pollution in the area. An inventory is compiled by listing the source types, calculating how much of a given pollutant is likely to be produced by, for instance, burning unit mass of fuel in each process (the *emission factor*, see table 1.12), determining the number of each source type in the area and estimating the total areal emissions. Compiling such an inventory can be a lengthy and sometimes costly business. Emission factors are important and yet not well determined, varying by up to 50 % depending upon the researcher and existing parameter values at the time of the investigation. The factor may not even be constant. For example, CO emission from cars depends upon the overall efficiency and maintenance of the engine, the load and whether the car is idling, cruising or braking. An overall average factor is often taken, hence the relatively large errors inherent in an estimation of an emission factor.

Table 1.12   Some emission factors. (Single values of SO_2 used in the study of Cullis and Hirschler (1980); CO_2 data after Hirschler (1981).)

	SO_2 (kg per 1000 kg of product)	CO_2 (kg per kg of fuel)
Hard coal	48.2 (range about 5–60)	2.53
Lignite	35.6 (possibly up to 70)	1.55
Coke	5.4	2.28
Petroleum refining	2.0	
Petroleum		3.06
Fuel wood and forest fires		1.47
Copper smelting	2000	
Lead	470	
Pulp/paper	2	

SUMMARY

Although air pollution has had a long history, concern, expressed in terms of action, is recent. Success in ridding urban areas of high concentrations of smoke, grit and sulphur dioxide, wrought partially by the implementation of Clean Air legislation and partly by a trend in domestic heating away from coal (a trend which may be reversed towards the end of the century as oil and gas reserves dwindle) can be claimed for about the past thirty years. Today, with lower ambient concentrations, emphasis is being placed on the control of localised emissions, on global trends and on chemical species previously ignored (e.g. CFCs, heavy metals). Within this framework (applicable to the developed world) and bearing in mind a possible similar path for newly industrialising nations, the major pollutants have been delineated and basic source types identified. To illustrate the magnitude of the problem, emission (source) data from two adequately monitored countries (USA and UK) have been presented.

2
Meteorology and Climatology

Dispersion and transport of pollutants are governed primarily by the meteorological conditions prevailing at, and subsequent to, the time of emission. In this chapter the physics and chemistry of the atmosphere are outlined. The vertical structure of the atmosphere is important in determining the stability and hence the vertical dispersion characteristics. The stability or instability is related to one of two reference lines: the dry and the saturated adiabatic lapse rates (dependent upon whether the air is saturated with water vapour). A comparison between this reference and the actual environmental lapse rate determines if or where a rising buoyant pollutant will reach equilibrium. Stable layers (inversions) are notorious for creating poor dispersion conditions and high ground level concentrations. The discussion is then widened to include synoptic conditions associated with both low and high dispersion conditions. Since most pollutants are emitted into an urban environment it is necessary to investigate the effects of the urban fabric on the speed and direction of the wind on a variety of space scales. Long term effects may be of greater importance for the future evolution of the atmosphere. These are discussed with relation to increasing carbon dioxide (the possible increase in surface temperature resulting from an enhanced 'greenhouse' effect) and decreasing ozone concentrations (a possible increase in ultraviolet flux at ground level causing an increase in the incidence of skin cancers). Air pollution climatology is becoming an increasingly important subject as anthropogenic emissions of such pollutants may be accumulating sufficiently to have an impact on the global climate as well as local (especially urban) climates.

THE VERTICAL STRUCTURE OF THE ATMOSPHERE

Most of the Earth's atmosphere is near the ground. 99.99 % of the atmosphere is below a height of 80 km—the composition being roughly that of table 1.1. The decrease of density with height is largely a result of gravitational effects and hence there is some differentiation between molecular species in their rate of decrease. A plot of concentration against height for a single compound is usually found to be *exponential* with a *scale height* (the height at which the concentration is 10 % of that at ground level) characteristic of the species. Scale heights are of the order of 10 km.

There are some anomalies in the exponential distribution of atmospheric gases. Perhaps the most important of these is ozone, which has a peak in the stratosphere at about 50 km above the Earth's surface. The ozone layer is relatively thin but plays an important role in absorbing incoming ultraviolet radiation which would otherwise make advanced life forms (on the surface of the Earth as opposed to underwater) impossible. The radiation absorption properties of ozone lead to an increase in temperature in the stratosphere. Maxima and minima in temperature are used to divide the atmosphere into layers. These are depicted in figure 2.1. The lowest layer is the troposphere (from the Greek *tropos* meaning turning, i.e. mixing). This extends to a height of about 10 km, dependent upon the latitude. All weather occurs in the troposphere and almost all air pollution is confined to this region. The temperature gradient discontinuity, which is in effect an *inversion base*, at the top of the troposphere is the *tropopause*. Below this level temperature decreases with height; above the tropopause, in the stratosphere, temperature increases to a height of about 50 km and a temperature of about 270 K at the stratopause. The next layer is the mesosphere in which the temperature again decreases with increasing height up to about 80 km (where the temperature is approximately 180 K). Above that height, electrical phenomena dominate and, although the temperature of the outermost layer is several hundreds of degrees centigrade, there are very few molecules and the temperature is simply an indication of their high individual kinetic energies.

The above description of the vertical structure of the atmosphere is valid 'on average'. With the exception of supersonic aircraft flying (and emitting nitrogen oxides and water vapour) in the stratosphere, all our discussion of air pollution meteorology will be confined to the troposphere—whether on a small scale or on a global scale. The vertical temperature structure has a large part to play in either dampening or permitting turbulent mixing—and hence dilution and dispersion—of air pollutants. This is expressed by means of the *stability*, usually discussed in terms of *lapse rates*.

LAPSE RATES AND STABILITY

Imagine a 'parcel' or lump of air that is lifted *in toto* from ground level upwards. Higher in the atmosphere, the lower atmospheric pressure allows the air parcel to expand. If this is assumed to occur with no exchange of heat between parcel and surroundings (an *adiabatic* process) it is found that the temperature of the parcel decreases. For a dry parcel (i.e. containing no or little water vapour), the decrease of temperature occurs at a constant rate known as the *dry adiabatic lapse rate* (DALR).

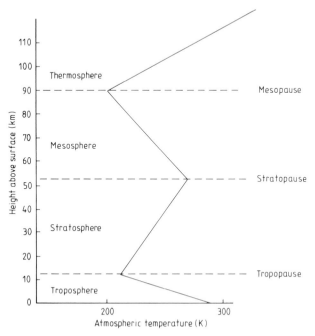

Figure 2.1 Average atmospheric temperature profile.

Mathematically the DALR is evaluated using the laws of thermodynamics and the ideal gas law (see equation 1.1). The first law of thermodynamics relates the energy change, dE, during a process, to the changes in temperature, dT, volume, dV, and pressure, p:

$$dE = c_v \, dT + p \, dV \tag{2.1}$$

where c_v is a constant known as the specific heat at constant volume. It has been stated that an adiabatic process is one in which no energy exchanges occur, i.e. d$E = 0$. Thus the equation for an adiabatic process is

$$c_v \, dT + p \, dV = 0. \tag{2.2}$$

To derive the lapse rate for this adiabatic process, the temperature must be derived as a function of height. To do this, the ideal gas law (equation 1.1) is used in its differential form

$$p \, dV + V \, dp = R \, dT \tag{2.3}$$

in order to eliminate dV, and in its ordinary form to eliminate V. Hence

$$(c_v + R) \, dT = (RT/p) \, dp. \tag{2.4}$$

$(c_v + R)$ can be shown to be equal to the specific heat at constant pressure, c_p. Furthermore, for a system in hydrostatic equilibrium, the barometric (or hydrostatic) equation holds, which gives

$$\frac{\mathrm{d}p}{\mathrm{d}z} = - \varrho g \qquad (2.5)$$

where ϱ is the density of air, g the acceleration due to gravity and $\mathrm{d}z$ the change in altitude. Thus $\mathrm{d}p$ is eliminated

$$c_p \, \mathrm{d}T = - \frac{RT}{p} \, \varrho g \, \mathrm{d}z$$

$$= - g \, \mathrm{d}z \quad \text{(since } p = \varrho RT\text{)}$$

therefore

$$\frac{\mathrm{d}T}{\mathrm{d}z} = - \frac{g}{c_p} \quad . \qquad (2.6)$$

This is the required dry adiabatic lapse rate. Both g and c_p have known constant values, the ratio of which is found to be -9.8 K km^{-1}. (Note: the word *lapse* implies a negative rate of change or decrease e.g. a lapse rate of 9.8 K km^{-1} is a rate of decrease of 9.8 K km^{-1}. Consequently a lapse rate which is *negative* corresponds to a rate of increase of temperature with height. This is the convention used in this book. The reader is advised to take care when reading the work of some other authors where lapse rates of $- 9.8 \text{ K km}^{-1}$ are related to a temperature decrease, the opposite meaning to that used here. The choice of convention could be considered as a matter of taste.) Thus taking a lapse rate of 9.8 K km^{-1} as a decrease of temperature with height allows us to specify the temperature change as

$$\frac{\mathrm{d}T}{\mathrm{d}z} = - \Gamma_d \qquad (2.7)$$

where Γ_d is the dry adiabatic lapse rate with a (positive) value of 9.8 K km^{-1}. The rate of temperature change is the negative of this. This convention is thus self-consistent and will be referred to later in this section in the discussion of stability.

Figure 2.2 illustrates the DALR, which is primarily a *reference* line used in discussion of stability. The actual rate at which the atmospheric temperature changes with height is the *environmental lapse rate* and is unlikely to be equivalent to the DALR. If, however, the environmental lapse rate is equivalent to the DALR the atmosphere is said to be *neutral* or in *neutral equilibrium*. The stability of the atmosphere is defined with reference to this neutrally stable state by further consideration of a hypothetical parcel of air. Consider an environmental lapse rate of 3 K km^{-1}, i.e. one where the temperature falls off less rapidly with height than the DALR (see figure 2.2). Suppose a parcel starts at ground level (surface temperature 288 K) and is lifted so that its physical characteristics (pressure, volume, temperature) change adiabatically. Then the rate of cooling is 9.8 K for every kilometre gain in height. If the parcel is raised to a height of 100 m its temperature will be $288 - 0.1(9.8) = 287.02$ K. However, at that height, the atmospheric temperature is $288 - 0.1(3) = 287.7$ K. The temperature of the parcel of air at 100 m is thus less than that of the surrounding air. Its density is thus greater and there is a net restoring force downwards (figure 2.2). This force exists until the parcel returns to its original position. Vertical motions will thus not occur spontaneously. The system is severely damped and the atmosphere is said to be *stable*. Thus in stable conditions the lapse rate (which is given by a positive number) is *less* than the DALR and the associated temperature gradient is *less negative* (i.e. *greater*). The problems of comparing two negative numbers should be borne in mind.

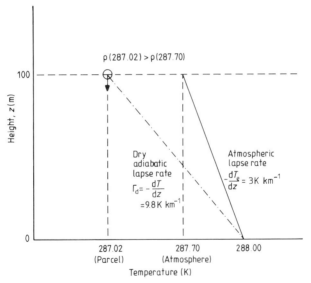

Figure 2.2 Stability of 'air parcels' for a lapse rate of 3 K km^{-1}.

Conversely if the environmental lapse rate is greater than the DALR a rising parcel will continue to rise, so that only a small disturbance is seen to be necessary to initiate vertical motion. This set of conditions represents instability. The unstable and stable regions are indicated on figure 2.3. The influence of stability on effluent dispersal is discussed in detail later in this chapter.

An alternative graphical display is shown in figure 2.4 using the concept of *potential temperature*. The DALR results directly from the fact that air is a compressible fluid and thus pressure changes with height even under neutral conditions. To overcome the problem of changing variables, it is observed that air at different heights in a neutral atmosphere (and thus at different temperatures) would possess the same temperature if brought down to the 10^5 N m^{-2} (1000 mbar) pressure level ('along the DALR'). This ground level temperature is the potential temperature, θ, defined by

$$\theta = T\left(\frac{p_s}{p}\right)^{R/c_p} \tag{2.8}$$

in which p_s is the surface pressure equal to 10^5 N m^{-2}.

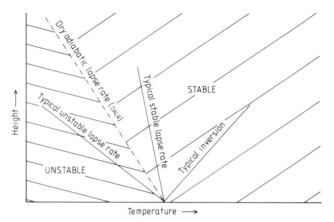

Figure 2.3 Stable and unstable regions for dry parcels.

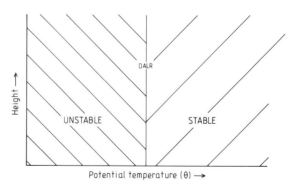

Figure 2.4 Stable and unstable regions defined in terms of the potential temperature θ.

By differentiating (and use of the ideal gas law etc) it is easily shown that the potential temperature gradient $d\theta/dz$ is related to the actual temperature gradient dT/dz by

$$\frac{1}{\theta}\frac{d\theta}{dz} = \frac{1}{T}\left(\frac{dT}{dz} + \frac{g}{c_p}\right). \tag{2.9}$$

Noting that in a neutral atmosphere $dT/dz = - \Gamma_d = g/c_p$, then $d\theta/dz = 0$. Hence the vertical line in figure 2.4 ($d\theta/dz = 0$) indicates neutrality. Positive slopes (of $d\theta/dz$) are stable, negative slopes unstable. The use of θ thus assists in clarifying discussions of stability†.

There is however one exception. An *inversion* is defined as being the situation when the atmospheric lapse rate is negative (i.e. the temperature gradient is positive and the environmental temperature *increases* with height); there is no corresponding definition (in common usage) in terms of potential temperature. An inversion is always stable; but stable conditions are not always inversions. Inversions are effectively barriers to vertical motions and have a strong damping effect on dispersion of pollutants in the vertical direction (see discussion on dispersion in Chapter 3).

Radiation inversions

The normal temperature profile is one in which there is a positive lapse rate (i.e. the temperature decreases with increasing height). Cases in which the temperature increases with height (negative lapse rate), even if only over a shallow depth, are called inversions and contain air of *high stability*. Their occurrence may be due to one of several reasons. Perhaps the most common is the *low level radiation inversion*. This tends to form overnight under conditions of light wind and clear skies. At the end of a warm day the environmental lapse rate will be approximately adiabatic since mixing processes active during the day ensure the atmosphere is homogeneous. During the hours of darkness the Earth loses heat by radiation and the layer of air next to the ground is cooled by conduction. A ground level inversion develops (see figure 2.5) by, say, 22.00 hours and is strengthened by about 02.00 hours the next morning. Cooling occurs at all levels throughout the early hours of the morning, but the inversion remains. Such conditions are often obvious soon after industries commence production in the early morning and effluent is trapped in the low level inversion layer (see figure 2.6). In the morning, after sunrise, the ground and the air above it begin to warm. The inversion is eroded (see figure 2.5) until at about midday a homogeneous, neutral atmosphere is once more attained.

Subsidence inversion

The second major mechanism for inversion formation is the prevalence of anticyclonic conditions which favour large-scale subsidence. As air subsides it warms, the upper parts of the subsidence layer descending (and thus warming) more than the lower layer. This forms a high level *subsidence inversion*. Subsidence inversions, once created, can remain for periods of several days, associated as they are with anticyclonic conditions, and may thus be an important factor in air pollution episodes.

†A further method of classifying stability is in terms of the Brunt-Väisälä frequency $N^2| = (g/\theta)\ d\theta/dz$; where again positive values of N^2 indicate stability and vice versa.

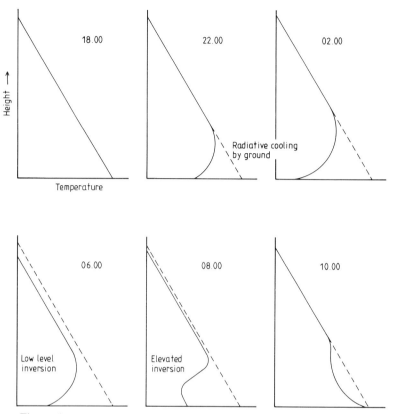

Figure 2.5 Stages in the formation of a low level inversion overnight.

Figure 2.6 An early morning inversion restricts plume rise and traps pollutants near ground level.

Other types of inversion

(a) *Frontal.* Frontal inversions are by their nature both transitory and not a significant air pollution dispersion problem as it is usual for frontal systems to have relatively high windspeeds associated with them. They occur because warm air is aloft and thus a temperature plot vertically through the system shows a discontinuity (see figure 2.7) and hence a shallow inversion layer.

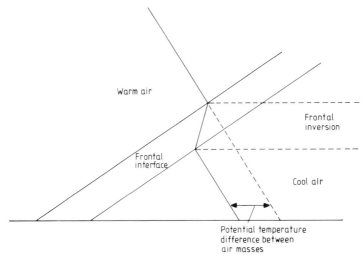

Figure 2.7 Frontal inversion.

(b) *Sea breezes.* A sea breeze circulation results in cool air near the ground and warmer air (rising above the land) overlying it. Thus, analogous to the frontal inversion, a weak inversion region exists embedded in the sea breeze circulation pattern.

The transitory and localised nature of both frontal and sea breeze inversions renders them of little importance in air pollution meteorology (although the localised closed circulation of a sea breeze may result in local recirculation of pollutants emitted within the cell, augmenting ground level concentrations as the day progresses).

Saturated adiabatic lapse rate

The above discussion assumes that there is only a small water vapour content in the atmosphere (i.e. less than the saturated value). For larger amounts it may be necessary to take into account latent heat released by condensing water vapour. As air cools, the maximum amount of water it can hold (when it is said to be 100% saturated) decreases. It is likely therefore that rising air may cool sufficiently to reach 100 % saturation. At a given height the degree of saturation is given by m/m_s where m is the mixing ratio and $m_s(z)$ the saturation mixing ratio (a monotonic decreasing function). The height at which this occurs depends largely on the value of m/m_s at ground level, namely the degree of saturation at ground level. If $m = m_s$, then water vapour will condense. Droplets of liquid water are seen as a cloud. Condensing water releases (latent) heat at a rate of about 2.5×10^6 J kg^{-1} (there is some dependence on temperature). Thus the rising air, which had been cooling adiabatically, is warmed by this heat release and cools down less rapidly if it continues to rise. Its rate of temperature decrease is given by the saturated adiabatic lapse rate (SALR) shown in figure 2.8. As more and more water is lost from the rising air, the rate of condensation slows—hence at great heights the SALR tends towards the DALR. Construction of the smooth curve for the SALR can be visualised in terms of linear 'infinitesimals'. As the air parcel rises an incremental distance, dz, it cools adiabatically and its temperature falls to T_A (figure 2.8). Since it is saturated, water condenses out and releases heat. The temperature thus rises to T_B. Another incremental rise, dz results in an adiabatic cooling to T_C followed by a release of latent heat and an increase in temperature to T_D. The SALR is then simply the smooth curve through T_o T_B T_D .... The equation of this line can be derived from the first law of thermodynamics. To the equation for dry rise (equation 2.1) must be added an additional term for latent heat transfers

$$- L \, dr_w = c_p \, dT - V \, dp \qquad (2.10)$$

where r_w is the saturation mixing ratio. Following similar arguments it is easily shown that the lapse rate is given by

$$-\frac{dT}{dz} = \Gamma_d + \frac{L}{c_p} \frac{dr_w}{dz} = \Gamma_s \qquad (2.11)$$

where Γ_s is the SALR which varies between about 5.0 and 9.8 K km^{-1}. Since the saturation mixing ratio decreases with height $dr/dz < 0$. Hence $\Gamma_s < \Gamma_d$.

Stability for a wet parcel is defined by reference to the SALR in a similar way to that for a dry parcel, i.e. in figure 2.9 environmental lapse rates to the right of the SALR are stable, to the left of the SALR unstable. Combining figure 2.9 with figure 2.4 defines three regions shown in figure 2.10. Region A is unstable and region C is stable with respect to wet and dry air parcels; whereas an atmospheric lapse rate falling in region B between the two lapse rates is found to be unstable with respect to wet parcels and stable with respect to dry parcels. This marginal behaviour is important in determining trajectories of saturated plumes (see Chapter 3) where the plume may alternately condense (wet plume) and re-evaporate (dry plume).

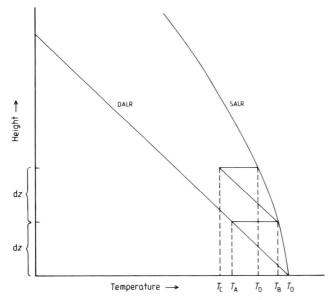

Figure 2.8 Derivation of saturated adiabatic lapse rate.

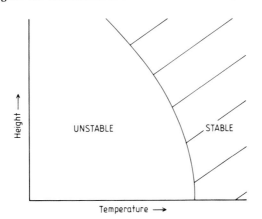

Figure 2.9 Stable and unstable regions for wet parcels (cf figure 2.3 for dry parcels).

AIR POLLUTION METEOROLOGY

In addition to the atmospheric temperature, the pressure patterns, strength of the wind and precipitation are also important for air pollution. It is perhaps worthwhile to give here a general introduction to synoptic meteorology.

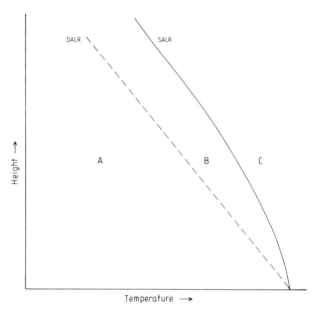

Figure 2.10 Stability classes for wet and dry plumes. A, wet unstable, dry unstable; B, wet unstable, dry stable; C, wet stable, dry stable.

In mid latitudes the weather is dominated by frontal systems. These form as a result of high level vorticity which induces cold air from polar regions to meet with warmer air flowing from tropical areas. The encounter between two air masses with contrasting temperatures delineates a line of discontinuity which is known as a *front*. Disturbances along this front (along which there is a wind shear of π radians (180°) (figure 2.11)) result in the line of the front beginning to 'bend' (see figure 2.12). The direction of travel of this developing frontal system is determined by the high level (about 500 mbar or 50 km high) winds which are roughly in the same direction as the winds in the *warm sector* at ground level. A cross section through this *warm sector depression* (figure 2.13) shows that warm air is forced to rise over a wedge of cooler, denser air along the lines of the fronts. Since this warmer air is often moist, it cools as it rises and the water vapour held in the air mass begins to condense, forming clouds. Thus along the fronts (the warm front preceeding the cold front) rain belts exist. As can be seen from figure 2.13, an observer on the ground perceives the rainbelt approaching him before the air mass at or near ground level changes from cool to warm. The reverse holds for the cold front—the rain starts after the change from warm to cool temperatures has been noted. In a fully developed depression (figure 2.14) not only frontal positions but also pressure patterns can be identified. From figure 2.13 it can be seen that, as a result of the two air masses being of a different temperature and hence a different density, the pressure (which is only a vertical integration of the air density, or a measure of the 'weight' of air pressing down on a given position) changes across a depression. Detailed consideration of atmospheric physics would show how these pressure patterns develop. When considering air pollution it is important to be able to identify regions of rain (for its cleansing properties) and regions of high wind. Winds are simply a result of pressure differences and hence related to the pressure patterns shown in figure 2.14. The pattern is described by a set of *isobars* which are lines joining places of equal pressure. The closer the isobars, the larger is the pressure difference (or *gradient*) and the stronger the wind-speed. Winds are discussed in more detail below.

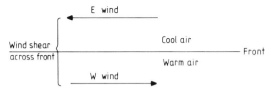

Figure 2.11 Wind shear across a frontal discontinuity when two air masses meet.

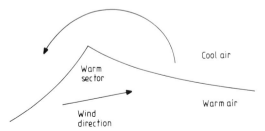

Figure 2.12 Wind directions and air masses associated with a developing depression system.

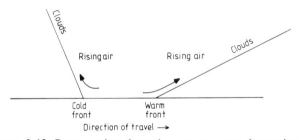

Figure 2.13 Cross section through a warm sector depression.

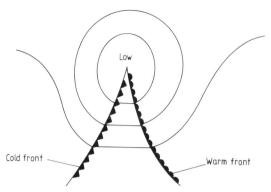

Figure 2.14 Plan view of a developed warm sector depression illustrating fronts. The rain bands precede the warm front and follow the cold front.

The two large-scale circulation patterns observable on a ground level synoptic map are depressions (cyclonic or anticlockwise flow in the northern hemisphere) and anticyclonic (clockwise flow in the northern hemisphere). Depressions (called cyclones or extra-tropical cyclones in the USA) bring relatively high winds, precipitation and changing air masses (and often visibility) and can usually be regarded as aiding dispersion of air pollutants. Anticyclones are, on the other hand, slow moving and persistent weather conditions. Such high pressure areas can remain stationary for days or weeks (as in the summer of 1976 in England and Wales). Anticyclones often give rise to what are called *stagnation conditions* and have been responsible for many major air pollution episodes worldwide, exacerbating existing pollutant levels by preventing upward diffusion.

WINDS

Winds are created by pressure differences and their strength (velocity) is proportional to the pressure gradient, and hence inversely proportional to the isobaric spacing on a synoptic map. If ground friction could be ignored, a balance would soon be reached between the pressure gradient force and the Coriolis force, which is proportional to the windspeed. This balance determines the wind velocity and the direction—parallel to the isobars—the so-called *geostrophic wind*. (This is actually the wind arising from a balance between the pressure gradient force and the Coriolis force only i.e. friction is assumed to be absent.) Wind nomenclature relates to the direction from which the wind blows. Friction has the effect of rotating the wind anti-clockwise (backing) and reducing the wind velocity to approximately 60 % of the geostrophic velocity over land (about 80 % over the sea).

This 'plan view' of the wind relates to geostrophic winds at the level at which surface friction becomes negligible. Hence close to the surface the wind may be of a different velocity and/or direction (see for example figure 2.15). Vertical variations may not be quite so dramatic as this wind shear— an 'average' vertical variation is often described by the logarithmic profile (figure 2.16)

$$U = a \ln(z/z_0) \tag{2.12}$$

in which U is the wind velocity at heigh z†, a is a function of the stability and z_0 is a constant dependent upon the roughness characteristics of the underlying surface. The value of this *roughness length* is very approximately 10 % of the physical obstruction size.

Figure 2.15 Wind shear can alter the direction of the plume dramatically as seen here in the Fiddler's Ferry power station cooling tower plume (UK).

Wind roses
A commonly used method of displaying wind data is the wind rose (figure 2.17). This is a polar diagram in which the direction of the wind is indicated by a bar in the direction *from which* it is blowing and the length of the bar indicates the percentage frequency of occurrence. This can be extended (figure 2.18) by evaluating the data into a discrete set of velocity classes. The width and/or shading of the bar thus indicates the velocity, whilst the frequency of occurrence and direction are as before. An alternative pictorial representation of windspeed is the boxplot (Paterson and Benjamin 1975, Graedel 1977). This is claimed to be a more useful representation although it is not yet in general use.

†Winds are usually measured at a height of either 2 or 10 metres.

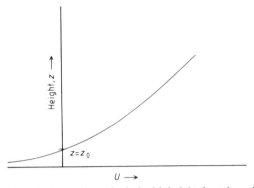

Figure 2.16 Vertical variation of wind with height for a logarithmic wind profile. The height z_0 is the roughness length and is where the windspeed (theoretically) falls to zero.

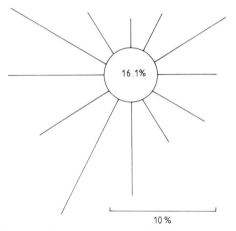

Figure 2.17 Mean frequency of wind direction at the Manchester Weather Centre (winds greater than 2.5 m s^{-1}). Data kindly supplied by the Manchester Weather Centre.

Windspeed varies not only as a result of pressure patterns but also because of ocean–atmosphere energy exchanges. Large releases of heat from the ocean into the atmosphere are aided by the autumnal equinoxial gales. This energy transfer is itself an input to the atmospheric flow (namely a positive feedback). On a diel scale (i.e. over 24 hours), windspeeds are in general higher in the afternoon than the pre-dawn hours of the morning.

In large urbanised areas, the pressure-induced wind patterns may be considerably disturbed. This disturbance can be viewed on a small scale, such as eddying flows around entrances to tall buildings, or on the city scale, where surface winds tend to flow towards the centre as a result of the *heat island* effect (see below).

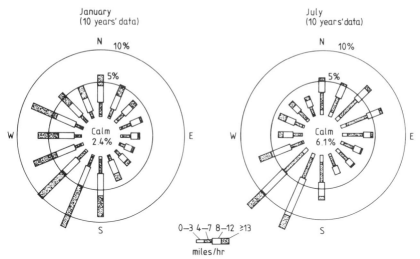

Figure 2.18 January and July wind roses, Cincinnati. The monthly distributions of wind direction and windspeed are summarised on polar diagrams. The positions of the spokes show the direction from which the wind was blowing, the length of the segments indicates the percentage of the speeds in various groups. Redrawn from Smith (1968) with the permission of the American Society of Mechanical Engineers.

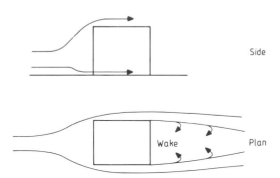

Figure 2.19 Wind flow around buildings.

Wind flow around buildings

In a rural, flat terrain, environment, the lateral wind field can be considered uniform. In an urban environment, however, many large obstacles, such as buildings, may be encountered, necessitating a deviation in the wind over or around the building (see figure 2.19). From continuity considerations it is seen that the deviated air adds to the total mass flow around the edges of the buildings. Since the mass flow rate must be unchanged, the velocity must increase. This phenomenon is readily observed around tall buildings on a windy day when the wind strength is higher around the building sides than about 20–50 metres away. However this increase in the wind velocity is not identical at all locations around the building. In the lee of the building there may be a stagnation point where there is no wind. This lies in the 'shielded' area downwind known as the *wake* (see figure 2.19). The wind shear across the boundary of the wake creates circulating *eddies* within the wake and this may result in a flow reversal somewhere within this region.

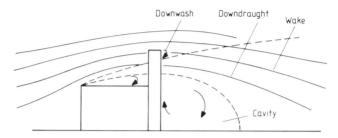

Figure 2.20 Downwash and downdraught effects shown schematically and as observed at Warrington, Cheshire (UK).

Figure 2.19 shows a plan view, but a similar effect is seen in all three dimensions (see also figure 2.20) and is an important concept in determining air pollutant concentrations. Pollutants may be trapped within the wake of the low pressure *cavity* and be prevented from dispersing upwards. This phenomenon is known as *downdraught*; the effect of *downwash* is a direct result of effluent being sucked down into the low pressure region in the lee of the chimney itself. A study of the chimney top can indicate prevailing wind direction since the lee of the chimney is likely to exhibit darkening due to emitted pollutants. The occurrence of downwash is governed by the ratio A

$$A = w_0/U \tag{2.13}$$

where w_0 is the exit velocity of the effluent and U is the windspeed at the stack top. Downwash is avoided if $A > 1.5$ but occurrence is unavoidable if $A < 1.0$. There is a range of ambiguous values for $1.0 \leqslant A \leqslant 1.5$. These criteria are based largely on wind tunnel studies of momentum-dominated effluents. Buoyancy effects can assist in overcoming downwash—this is shown to be the case if the Froude number (given by the ratio of momentum forces to buoyancy forces) is less than or equal to 1.0.

Usually the exit velocity is determined by the physical dimensions of the chimney, the boiler capacity and the gas flow rate through the chimney, although exit velocities can be increased by using the venturi effect† (see figure 2.21). A range of windspeeds for which downwash is likely can thus be calculated if the effluent exit velocity is known, namely

$$U \geqslant w_0/1.5. \tag{2.14}$$

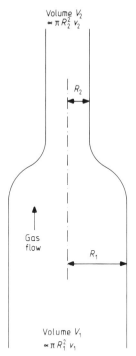

Figure 2.21 The venturi effect. Mass continuity gives $\pi R_1{}^2 \, v_1 = \pi R_2{}^2 \, v_2$. Hence $v_2/v_1 = (R_1/R_2)^2$.

---

†If A becomes too large, acid droplets become entrained into the waste gases. The problem of these 'carryover' particles can be of great importance and such emissions should be minimised.

Thus avoidance of downwash is technically possible although in practice it can be a relatively frequent occurrence; an example is shown in figure 2.22.

Downdraught effects are avoided if pollutants are emitted from a sufficiently tall chimney. As a rule of thumb, the chimney should be at least 2.5 times the maximum height of neighbouring buildings. For tall, thin buildings this is over conservative and the height is determined (Snyder and Lawson 1976) by $h + 1.5\ l$ (where h is the building height and l is the smallest of the maximum building widths perpendicular to the wind direction and the value of h).

Figure 2.22 The Fall River (Massachusetts, USA) incinerator plume, exhibiting stack tip downwash.

HEAT ISLANDS

On a larger scale, the wind patterns within an urban area may also show flow patterns disturbed from the overall forcing synoptic situation. The urban temperature is increased as a result of the large amount of heat released from the urban fabric, due to both industrial waste heat and domestic and commercial space heating, together with decreased heat loss associated with lower availability of water for evaporation and with reduced windspeeds caused by high values of surface roughness. The greatest increase is in the most developed area (usually the centre) so that a horizontal temperature gradient is established. At the simplest level this can be described in terms of the relatively warm air at the centre which rises, drawing in air from the suburbs (figure 2.23). The rising air eventually ceases to rise and flows radially outwards at a height of a few hundred metres, slowly descending to replace the ground level air in the suburbs thus completing the circulation. When this heat island has developed, any pollution emitted anywhere within this closed system will simply recirculate and remain trapped within the heat dome. Pollutant concentrations build up, although the values may be moderated by the vertical migration of the top of the mixed layer, which often acts as a lid in the same way as an inversion. For a more detailed discussion of these effects see Chapter 3.

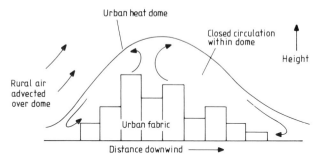

Figure 2.23 Closed circulation within the urban heat dome.

AIR POLLUTION CLIMATOLOGY

In Chapter 1, average atmospheric retention times for specific pollutants were discussed. Interactions between pollutant and atmosphere depend largely on the time available for interaction. Most pollutants are removed from the atmosphere locally or before they have had the opportunity to spread homogeneously across the globe. Some waste emissions may remain in the atmosphere long enough to affect the global background. The most important of these is carbon dioxide (CO_2), a product of complete combustion and thus anticipated to be a major end product of any combustion process. At present well over 12 000 million tonnes of CO_2 (from anthropogenic sources) are emitted into the atmosphere each year. Only about 50 % of this remains in the atmosphere yet this is sufficient to cause a notable increase in CO_2 concentration.

The enhanced global CO_2 concentration has become a cause of concern over the last 10–15 years as a result of the property of radiation absorption possessed by the CO_2 molecule. This can best be understood by a brief look at the radiation balance of the Earth–atmosphere system.

All energy received by the Earth comes from the Sun, which can be considered as a black body radiating at a temperature of about 5800 K. Black-body radiation results in a wavelength distribution of the available energy (figure 2.24), with the peak strength and wavelength being dependent solely on the effective temperature of the emitter. This distribution is given by Planck's law

$$E(\lambda) = \pi \, B_\lambda(T) = \frac{2\pi hc^2}{\lambda^5} \frac{1}{[\exp{(hc/\lambda kT)} - 1]} \tag{2.15}$$

where c is the velocity of light (2.998×10^8 ms^{-1}), k is Boltzmann's constant (1.3805×10^{-23} J K^{-1} molecule^{-1}), h is Planck's constant (6.626×10^{-34} J s) and λ is the wavelength. The maximum is given by Wien's law

$$\lambda_{\text{max}} = \frac{2900}{T} \ \mu\text{m}. \tag{2.16}$$

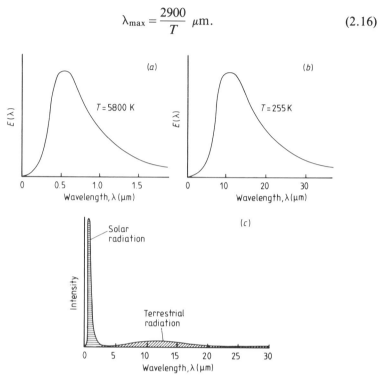

Figure 2.24 Variation with wavelength of energy emitted from a black body. Curve (a) corresponds to the Sun with emission temperature 5800 K and curve (b) to the Earth ($T = 255$ K). Both these curves are normalised. Curve (c) superimposes (a) and (b) on the same vertical axis.

The total energy flux (i.e. integrated over all wavelengths) is proportional to the fourth power of the absolute temperature. This is given by the Stefan-Boltzmann law

$$F = \epsilon\sigma T^4 \tag{2.17}$$

where σ is the Stefan-Boltzmann constant (5.67×10^{-8} W m^{-2} K^{-4}) and ϵ is the emissivity—a ratio measure of how well the emitter approaches the 'perfect' state. For a perfect black body $\epsilon = 1$. The solar flux received at the Earth can be calculated as approximately 1367 W m^{-2} of the solar beam, known as the solar constant S. The total energy intercepted by the Earth (which presents a cross section of πr^2, where r is the radius of the Earth) is thus $S\pi r^2$ W. Part of this is reflected away immediately. The ratio of reflected to incident energy is known as the *albedo, A* (from *albus* meaning white). Defined in this way it represents a value integrated over all wavelengths. The total solar radiation taken in by the Earth–atmosphere system is thus $S\pi r^2 (1 - A)$.

On the other hand the Earth itself radiates as a black body; thus losing a total energy of σT_e^4 W m^{-2} (where T_e is the *effective temperature* of the Earth). Assuming, as is the case for the present day Earth, that the whole of the spherical Earth–atmosphere system $(4\pi r^2)$ is involved in this radiation, the energy loss is $4\pi r^2 \sigma T_e^4$ W. For the Earth to be in stable thermal equilibrium the solar radiation input and the total energy loss must be equal, hence

$$S\pi r^2(1 - A) = 4\pi r^2 \sigma T_e^4 \tag{2.18}$$

or

$$S(1 - A) = 4\sigma T_e^4. \tag{2.19}$$

It is assumed that the emissivity of the Earth is unity. Solution of this equation gives a value for T_e of about 255 K—which is obviously not the present surface temperature. However, if there were no atmosphere, the surface temperature would indeed be 255 K ($-18\,^\circ$C). The radiation from the Earth's surface is at longer wavelengths than the incoming solar radiation. From Wien's law, the peak emission is found to occur at around 11 μm (see figure 2.24(b)). The range of terrestrial radiation lies in the infrared, from about 4 μm upwards. Fortuitously the amount of solar radiation at such wavelengths is negligible and the two Planck curves superimposed (figure 2.24(c)) can be described by a 'two-stream', effect. Atmospheric gases absorb at specific wavelengths—some in the solar range, some in the terrestrial. The two-stream effect permits separate consideration of the incident (shortwave) and emitted (longwave) fluxes. The main absorbers are water vapour, carbon dioxide and ozone. In the short wavelength region, ozone absorbs incoming ultraviolet radiation which would be lethal if it reached the Earth's surface unattenuated. On the other hand the largest absorption bands of CO_2 and H_2O lie in the infrared (see table 2.1).

TABLE 2.1 Major absorption bands longwave of 4 μm. The most important bands are italicised.

Gaseous absorber	Wavelengths (μm)		
O_3	4.7	10.0	15.0
CO_2	*4.3*	4.8	*15.0*
H_2O	6.3	>30.0	
CH_4	7.7		
N_2O	4.0	4.5	8.0
CO	4.7		

Longwave radiation absorbed by the atmosphere is eventually re-radiated (again in the infrared wavelengths corresponding to the Planck distribution associated with the atmospheric temperature at the height of emission). Re-radiation occurs in all directions and thus a portion of this radiation is re-received at the surface (see figure 2.25). The received radiation is thus greater than that given by the left-hand side of equation (2.18) and thus the surface temperature T_s is increased. This is usually expressed as

$$T_s = T_e + \Delta T \qquad (2.20)$$

where the increment due to the blanketing effect of the atmosphere is known as the *greenhouse increment*. The term 'greenhouse effect' is in fact a misnomer, resulting from the analogy that greenhouse glass permits shortwave (solar) radiation to pass through but absorbs outgoing thermal infrared. In fact recent analysis has shown that greenhouses heat up due to their enclosure of air which is prevented from cooling by the normal processes of convection or advection. The greenhouse increment on Earth depends upon the concentrations of the absorbing gases especially CO_2 and H_2O. Addition of CO_2, for example, to the atmosphere broadens the absorption bands and the greenhouse increment is increased. Over the years many climatologists have attempted to calculate the additional greenhouse increment which would result from a doubling of our present level of CO_2. The consensus is now being reached that a doubling should result in an increase in mean global temperature of 3 ± 1.5 K. However it is important to note that this is an average value and it now seems that some latitudes will suffer a considerably greater increase than the mean (of the order of 9–10 K in the polar regions).

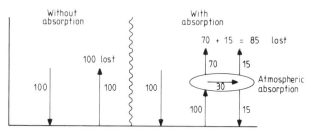

Figure 2.25 Incoming and outgoing radiation streams without and with atmospheric absorption (aerosols plus clouds) illustrating the additional surface heating in the latter case.

If fossil fuel combustion continues at its present exponential rate, the CO_2 concentration in the atmosphere will have doubled in about 50 years. However this assumption may well be an overestimate when the finite nature of fossil fuel reserves is recalled—although the extent of a switch to nuclear power, conservation policies, rises in oil prices and development of alternative energy sources should also be borne in mind. The effects of increased

CO_2 concentration could be wide ranging. The global general circulation patterns would most likely be changed, irrespective of sea level changes as a result of ice melt (which would be negligble for melting of North Polar ice, since this is pack ice and not terrestrial as it is in Antarctica). Such changes will alter rainfall patterns, and lengths of growing seasons which may make subsistence agriculture in certain areas untenable.

No one knows the direction or extent of climatological feedback. There is certainly a large degree of buffering by the oceans which absorb CO_2, ameliorate temperature changes and perform meridional transports of energy. The impending change from an oil to a coal economy decided upon by the European Summit in Venice in June 1980 has already been criticised as taking us back to the pea souper (sulphurous) smogs of the 1950s. Certainly changes in fuel type have long lasting implications for air pollution and pollution control technology will need to be developed appropriately. In addition larger usage of nuclear fuels will help slow down the carbon dioxide build up.

It is not only CO_2 that has threatened to pollute and change the global climate. In the 1960s there was great concern about the possible reduction in the ozone layer caused by emissions (largely NO_x and H_2O) from supersonic aircraft, such as Concorde, flying in the stratosphere. The fuel crisis and political pressure, together with the lobby against noise, have probably restricted use of such planes to a number well below that calculated as being detrimental to the ozone layer.

Aerosol spray cans have used 'inert' chlorofluorocarbons (CFC) as propellants. When it was discovered that these found their way into the stratosphere, again depleting the ozone layer, pressure was put on governments to 'ban the can'. Although CFCs have increased and have entered the stratosphere the chemistry is so complex that it is not yet clear whether the net effect in the ozone level will be positive or negative—self-regulating feedbacks may be present—see, for example, Lovelock's *Gaia Hypothesis* (1979). A National Academy of Sciences report in 1982 suggested that the decrease in ozone concentration by the end of the next century could be less (5–9 %) than previously forecast (15–18 %). In perspective any decrease in ozone concentration which might be likely to occur will cause a local increase, at ground level, of ultraviolet of smaller magnitude than would be observed by a traveller going from England to southern Spain or northern Africa. An increase in the incidence of skin cancer would be likely—yet many such cancers are non-fatal.

SUMMARY

The dynamics and chemistry of the atmosphere determine the efficiency of dispersion of pollutants emitted into it. This chapter has identified the following key concepts in air pollution meteorology and climatology which will be elaborated in Chapters 3,4 and 7.

Vertical temperature profile of the atmosphere.

Troposphere; lapse rates (DALR, SALR).

Stability—neutral, stable and unstable conditions.

Inversion formation.

Warm sector depressions and anticyclones.

Winds: log profile, wind roses, effects of buildings, downwash and downdraughts.

Urban meteorology: heat islands.

Climatology: radiation balance, greenhouse increment, effect of increasing CO_2 and depletion of ozone layer.

3

Dispersion and Plume Rise

The disposal of gaseous waste products is commonly undertaken by emitting them into the atmosphere. Once in the atmosphere the gases are dispersed by a variety of mixing mechanisms which dilute the effluents and transport them away from the point of emission. Although the effective dispersal relies on the atmosphere acting as an 'infinite sink', all atmospheric pollutants will eventually be removed from the atmosphere and returned to earth. The concentration of pollutants, measured at ground level, is the major criterion in assessing the efficiency of the disposal mechanism.

If a source of gaseous pollution is introduced into a still atmosphere, it begins to disperse due to the random motions of the gas molecules themselves. This process of molecular diffusion is well understood and easily predicted. The concentration, C, resulting from the pollutant release as it is dispersed away from the source, depends inversely on the distance x from the source such that $C \propto x^{-n}$ where n has a value between 1 and 2 (dependent upon the stability, see Chapter 2). Even when there is a wind, this relationship between concentration and distance appears to hold. Thus, the first attempt by engineers to reduce ground level concentrations from single sources was to emit the effluent into the atmosphere from a great height so that by the time the gases diffused down to ground level the pollutant concentration would be lowered. Thus the tall chimney policy was born, whereby the large height of the chimney not only creates sufficient draught, but also avoids downdraught effects (see Chapter 2, especially figure 2.20) in addition to reducing ground level concentrations.

It was then observed that if the waste gas was emitted from the chimney with an initial velocity, or if the gas was hotter than the ambient air, it would rise above the stack top before diffusing earthward. The height at which the effluent ceases to rise (the equilibrium height) led to the concept of effective stack height. This replaces the actual stack height in diffusion equations. Unfortunately, prediction of the height to which the effluent will rise is extremely difficult and highly dependent upon the meteorological conditions, as well as upon the momentum and buoyancy of the effluent.

The state-of-the-art was summarised by Pasquill (1971): 'This is a subject with a very wide and scattered literature, and with more formulae and dispute than in any other aspect of our present subject'.

PLUMES

Most chimneys emit effluent gases continuously. These are removed from the vicinity of the stack by a combination of the wind (translation), buoyancy plus vertical momentum (plume rise) and lateral diffusion. As the pollutant leaves the stack it forms a continuously expanding *plume* of material. The plume is blown downwind and at the same time may be displaced vertically, either up or down, by random fluctuations in the wind—these in addition to any vertical motion due to plume rise. An instantaneous photograph (see figure 3.1) shows a sinuous path for the plume and a 'lumpy' appearance. The path taken by the plume varies with time (figure 3.2 depicts the same plume as in figure 3.1 but photographed some seconds later). Prediction of this path cannot (at present) be undertaken since the four-dimensional wind field is not well understood. However, if a time exposure photograph of the plume is taken, then these irregularities in the plume path are smoothed and a conical or quasi-conical outline to the plume can be seen (see figure 3.3). It is this mean behaviour of the plume about which theories are formulated. For the purposes of prediction, meteorological parameters, such as windspeed, are usually approximated either as constants or as smoothly changing functions with time.

Figure 3.1 Instantaneous plume from European coastal industrial complex illustrating its 'ragged' appearance.

If a cross section of the plume is examined, then the values of temperature, density, velocity etc, are found to be dependent on the radial position (for a radially symmetric cross section). Maximum values usually occur at the plume centre, whilst near the periphery values approach those of the atmosphere itself.

Figure 3.2 The same plume as in figure 3.1 photographed some seconds later to illustrate small timescale fluctuations observed under otherwise steady conditions.

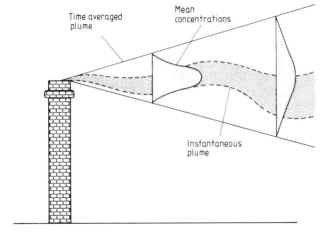

Figure 3.3 Diagrammatic representation of an instantaneous plume compared to the time-averaged plume and a laboratory simulation of the time-averaged plume.

The time-averaged profile of the plume has been shown experimentally to be well described by the Gaussian profile, centred on the mean axis of the plume. The instantaneous plume however is probably better described by the much used 'top hat' profile, relative to the instantaneous axis, where the plume parameters have one constant value inside the plume, another outside (this value varying with height). As time-averaged plumes will be described here, it would appear to be necessary to use the Gaussian profile, but it is easily shown that the top hat profile (with its simpler mathematics) gives identical results if 'model equivalent' parameters are used (Henderson-Sellers 1981).

Plume behaviour

Plume behaviour can be considered in two parts. The *plume rise* is a function of the heat content and efflux velocity of the plume governed by the atmospheric stratification (see Chapter 2) that may be present. When the plume reaches an equilibrium height (which may be at the stack height if the effluent is non-buoyant), the process of *diffusion* begins to dominate as a result of mixing by turbulent eddies in the atmosphere. (It should be noted that diffusion also occurs throughout the rising phase. However it is so much less effective than the dispersion by turbulent mixing which is induced by the passage of the plume through the atmosphere that it can be neglected at this stage. The section on plume rise will elaborate on diffusion during rising.) Diffusion is well understood theoretically, especially when the underlying terrain is flat, so a description of the process is worth studying. The case of a non-buoyant plume, emitted with zero velocity into a neutral atmosphere is used for simplicity.

DIFFUSION

Effluent emitted slowly (i.e. with negligible efflux velocity) from a chimney will be blown downwind at the height of the stack top. For an emission at a constant rate of Q kg s^{-1}, the volume of air into which this effluent is absorbed is directly proportional to the windspeed. Since the total mass emitted per second is constant, the volumetric concentration of pollution in the atmosphere, C, is inversely proportional to the windspeed, U

$$C \propto \frac{1}{U}. \tag{3.1}$$

The full relationship is derived by solution of the diffusion equation

$$\frac{DC}{Dt} = \frac{\partial C}{\partial t} + \frac{u\partial C}{\partial x} + \frac{v\partial C}{\partial y} + \frac{w\partial C}{\partial z}$$

$$= K\nabla^2 C$$

$$= K\left[\frac{\partial^2 C}{\partial x^2} + \frac{\partial^2 C}{\partial y^2} + \frac{\partial^2 C}{\partial z^2}\right] \tag{3.2}$$

which is the direct analogue of Fick's law for molecular diffusion. In equation 3.2 (for turbulent diffusion) K is the coefficient for turbulent diffusion (which is several orders of magnitude greater than the coefficient for molecular diffusion). The equation applied to stack plumes is usually solved for the special case of the steady state for a continuous point source. The result, for concentration C at any point (x, y, z) is given by

$$C(x,y,z) = \frac{Q}{4\pi r \sqrt{K_y K_z}} \exp\left[-\frac{U}{4x}\left(\frac{y^2}{K_y} + \frac{z^2}{K_z}\right)\right] \qquad (3.3)$$

where K_y and K_z are the components of the eddy diffusion in the y and z directions and $r = \sqrt{x^2 + y^2 + z^2}$.

The continuous point source solution is well known and often quoted— the so-called *Gaussian plume model*. The origin of the name is seen by examining the crosswind profile: at a given distance downwind ($x = x_1$, say), a cross section vertically through the axis of the plume (at $y = 0$) gives

$$C(x_1,0,z) = \frac{Q}{4\pi r \sqrt{K_y K_z}} \exp\left(-\frac{U}{4x_1}\frac{z^2}{K_z}\right) \qquad (3.4)$$

which is a Gaussian or normal form. Similarly a horizontal cross section ($z = \text{constant} = z_1$) is given by

$$C(x_1,y,z_1) \propto \exp\left(-\frac{U}{4x_1}\frac{y^2}{K_y}\right). \qquad (3.5)$$

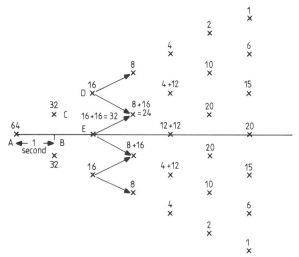

Figure 3.4 Cross-wind spread of 64 units of pollution over a time interval of 6 seconds.

The form of the distribution can be demonstrated simply as follows. A source of 64 units of pollution is released from rest at point A in a constant wind of 2 m s^{-1} (see figure 3.4 which gives a plan view). Diffusion processes operate and spread the plume material sideways. There is no preference for any direction and thus an equal amount of pollutant diffuses to the left and to the right. Let us assume that this diffusion occurs in one second (point B) and spreads the plume to a maximum of one unit distance to each side. The same process occurs in the next second. Half of the pollution at point C goes to point D and half to point E, which is on the centre line axis once more. The process continues. After six seconds the concentration at the outermost edge of the plume is only one unit. Plotting the concentrations at different distances from the plume centre reveals the distinctive bell-shaped Gaussian distribution (see figure 3.5).

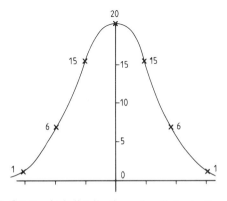

Figure 3.5 Cross-wind distribution of pollutant after 6 seconds.

This holds at any distance downwind. From equation 3.2 it is easily shown that

(i) the crosswind distribution ($z = 0$) is $a \exp(-by^2)$

(ii) the vertical profile ($y = 0$) is $c \exp(-dz^2)$

where a, b, c and d are constants, and thus the Gaussian distribution holds in all directions (see figure 3.6). The width of the plume thus depends upon the values of K_y and K_z. It is often assumed that these are equal (to K) in which case the plume is spherically symmetric. An alternative representation is often used, based on the statistical theory of turbulence developed largely by G I Taylor (1921). Instead of representing dispersion in terms of K, the standard deviations σ_y, and σ_z are used. The corresponding expression for the concentration is

$$C = \frac{Q}{2\pi\sigma_y\sigma_z U}\exp\left[-\frac{1}{2}\left(\frac{y^2}{\sigma_y^2} + \frac{z^2}{\sigma_z^2}\right)\right]. \tag{3.6}$$

Once again the distribution across the plume is found to be Gaussian.

Dependence on position downwind is not expressed explicitly; although σ_y and σ_z both depend on this distance and are given by

$$\sigma_y^2 = \frac{2K_y x}{U} \qquad \sigma_z^2 = \frac{2K_z x}{U}. \tag{3.7}$$

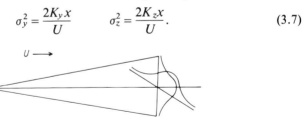

Figure 3.6 A Gaussian distribution holds across the plume cross section in any direction.

Dispersion parameters

The dispersion parameters σ_y, σ_z, discussed above, must be determined experimentally. Turner (1970) has summarised these results graphically as shown in figure 3.7. The curves obtained are often referred to as the Pasquill–Gifford (PG) curves. In these diagrams, values for σ_y (figure 3.7(a)) and σ_z (figure 3.7(b)) are depicted as a function of distance downwind and of the stability category. These stability categories are given in table 3.1 in which classes A, B and C are unstable, D is the neutral case and E, F and G are stable. Typical frequencies of occurrence for the UK are given in table 3.2. These σ values (correct to within a factor of two) are only valid for *surface* releases and sampling times of about three minutes in flat, open country. Furthermore these graphical values may be given by $\sigma = b_1 x^{b_2}$ or alternatively by the following expressions for σ_y and σ_z

$$\sigma_y = L_1 x / [1 + (x/l)^p] \tag{3.8}$$

$$\sigma_z = L_2 x / [1 + (x/l)^q] \tag{3.9}$$

where the values of b_1 and b_2 are functions of stability and position; the values of L_1, L_2, l, p and q are given in table 3.3 as functions of stability class.

In an urban environment, these values must be changed to take into account the excess turbulence generated by thermal and mechanical effects. Modified curves are presented by Briggs and Smith (in Miller 1978) for various roughnesses, and by Bowne (1974) for urban and rural cases. More recent observations in Canada were found to be in agreement with the original curves but suggested an additional Richardson number dependence. A review of the American Meteorological Society (AMS) Workshop by Hanna *et al* (1977) suggests the AMS curves are more appropriate for elevated releases, in addition to questioning the applicability of the Pasquill–Gifford curves, especially in unstable conditions.

Table 3.1 Key to stability categories used in figures 3.7 and 3.8. Classes A, B and C are unstable; E, F and G are stable. The neutral class, D, should be assumed for overcast conditions during day or night (after Turner 1970).

Surface wind-speed (at 10 m) (m s^{-1})	Day			Night	
	Incoming solar radiation			Thinly|overcast or	
	Strong	Moderate	Slight	$\geqslant \frac{4}{8}$ low cloud	$\leqslant \frac{3}{8}$ cloud
< 2	A	A–B	B		G
2–3	A–B	B	C	E	F
3–5	B	B–C	C	D	E
5–6	C	C–D	D	D	D
> 6	C	D	D	D	D

Table 3.2 Typical annual frequency of occurrence of stability categories given in table 3.1 (UK values).

Stability category	A	B	C	D	E	F	G
Frequency of occurrence (%)	0.6	6.0	17.0	60.0	7.0	8.0	1.4

Table 3.3 Dispersion parameters (after Green *et al* 1980).

	A	B	C	D	E	F
l(km)	0.927	0.370	0.283	0.707	1.07	1.17
L_2(m km^{-1})	102.0	96.2	72.2	47.5	33.5	22.0
q	−1.918	−0.101	0.102	0.465	0.624	0.700
L_1(m km^{-1})	250	202	134	78.7	56.6	37.0
p	0.189	0.162	0.134	0.135	0.137	0.134

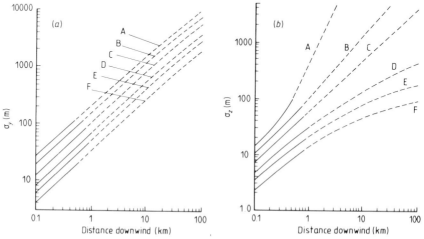

Figure 3.7 Dispersion coefficients as a function of downwind distance from the source. (*a*) Horizontal values (σ_y); (*b*) vertical values (σ_z). Redrawn from Turner (1970).

For a buoyant plume, an additional term of $\Delta H^2/10$ should be added to the value of σ_z^2 derived from the Pasquill–Gifford curves to give the real value of σ_z. The general uncertainty about the applicability of the PG curves is highlighted by many research studies in the past five years (e.g. Irwin 1983). At present, however, there appears to be no general consensus as to what curves should replace these since more variables may need to be incorporated in the calculation scheme. The interim scheme proposed following the AMS workshop relates the dispersion parameters to meteorological observations via a series of graphs and equations. Expressed in terms of maximum ground level concentration, C_{max}, graphs can be drawn as a function of surface roughness, z_0. For a typical urban value of $z_0 \simeq 1.0$ m, these interim results are shown in figure 3.8 in comparison with similar results derived from the Pasquill–Gifford curves of figure 3.7.

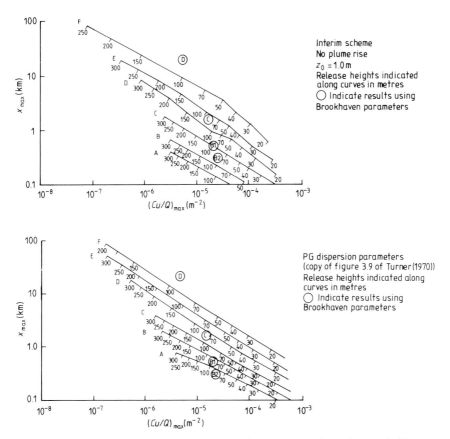

Figure 3.8 Distance of maximum concentration, C_{max}, and maximum Cu/Q as a function of stability (curves) and effective height (metres) of emission. Redrawn from Irwin (1979) with the permission of the American Meteorological Society.

Knowledge of the value of the vertical dispersion parameter is useful in determining where and when the plume will first touch the ground. Assuming a Gaussian profile, the concentration at a distance of 2.15 σ_z from the centre of the plume is 10% of the axial value and thus gives a reasonable estimate for the plume radius. The plume first touches ground when its radius is equal to the chimney height (or equilibrium height for a buoyant effluent), i.e. when

$$2.15\ \sigma_z = H. \tag{3.10}$$

As an example, for $H = 60$ metres, σ_z is found to be equal to 28 m. This value for σ_z is attained (see figure 3.7) at a distance of 0.7 km (stability category D) or 2 km (stability category F) downwind from the source. For $H = 120$ m, these distances are increased to 2 and 10.5 km respectively.

The Gaussian formulation can then be used to calculate the ground level concentration. To take into account the fact that the pollutant does not vanish at or 'below' ground level, the method of images is invoked (figure 3.9). Ground level concentrations over flat terrain are found to be twice the calculated value.

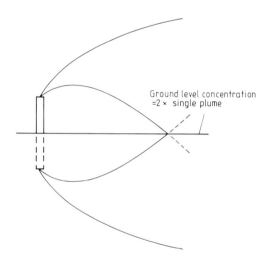

Ground level concentration
=2 × single plume

Figure 3.9 The method of images.

SOURCE STRENGTH

It has been assumed in the above discussion that the source strength is known in terms of its mass emission rate, Q_m (in kg s^{-1}). Alternatively some formulae require the emission rate in terms of its heat (or energy) content.

This can be given by Q_h (watts)

$$Q_h = Q_m c_p (T_0 - T_e) \tag{3.11}$$

where c_p is the specific heat at constant pressure in J kg^{-1} K^{-1}, T_0 the effluent temperature and T_e the ambient (environmental) temperature. The mass flux, Q_m, is given simply as

$$Q_m = \frac{\pi D^2}{4} w_0 \varrho_0 \tag{3.12}$$

where D is the stack diameter, ϱ_0 the density of the effluent and w_0 the exit velocity.

ATMOSPHERIC STABILITY AND CHARACTERISTIC PLUME SHAPES

A full discussion of atmospheric stability is to be found in Chapter 2. It is generally accepted that there are five basic plume behaviour types and these are shown in figure 3.10. When the atmosphere is neutrally stable, the symmetrical conical shape can be observed. Under inversion conditions, severe vertical damping is imposed on plume rise and vertical diffusion, although lateral spreading still occurs—the observed shape leads to the use of the term *fanning* as a description. If a two-layer stratification exists, then it is important whether the base of the inversion is above or below the chimney top. In the third case illustrated an inversion exists just above the chimney, restricting upward vertical spread. If the lower atmosphere is unstable, then rapid mixing down to ground level occurs, a process known as *fumigation*, and concentrations of pollutants rise. If a stable layer is capped by an unstable layer, then vertical mixing occurs only in the upper layer and pollution is trapped at a height greater than the chimney top—often known as *lofting*. If unstable conditions exist everywhere, then large random eddies are responsible for the *looping* conditions. Fumigation and looping, the two cases associated with an unstable environment in the lower atmosphere, are probably the most important.

Looping can occur whenever large scale eddies are present. These may be introduced by gusts in the wind, or by mechanical stirring if the airflow passes over some obstacle. The size of the eddies regulates the type of plume dispersion (see figure 3.11). Small scale turbulence will have little effect in changing the plume shape. Large eddies (compared with the plume radius) will tend to move the plume bodily but do little to enhance the growth in size so the plume meanders widely. In a mixed field of eddies of varying sizes the plume will both meander and expand as it is blown downwind.

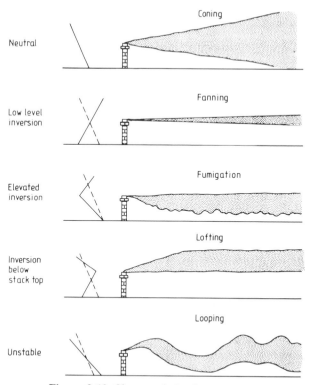

Figure 3.10 Characteristic plume types.

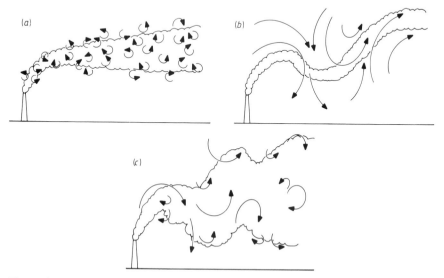

Figure 3.11 Plume dispersal in a field of eddies. Redrawn from Perkins (1974) by permission.

These shapes are observed in both passive plumes (e.g. as described by the Gaussian plume model) and also superposed upon rising plumes. However, the turbulence associated with the rising phase may obscure these characteristics such that only when the plume has reached its equilibrium level can these descriptions be applied. Since most effluents do indeed rise, either by virtue of their momentum or buoyancy, the subject of plume rise is now considered.

PLUME RISE

When effluent is released from a chimney it is unlikely that it will have a density equal to that of the environment, or that it will have no velocity of its own. These two important parameters (density difference or buoyancy and velocity) determine the plume behaviour. Although the word plume is used here to describe any continuous effluent, different terms are used in the literature to indicate the relative importance of momentum (mass times velocity) and buoyancy. If the effluent is released from rest, its motion is determined by its buoyancy, the result being a pure *plume*. If the effluent has no buoyancy but is ejected at a high velocity it is known as a *jet*. For effluents where both momentum and buoyancy are important the terms *buoyant jet* and *forced plume* are used interchangeably. In practice, many chimney effluents are emitted vertically upwards and the momentum (jet-like) effect dominates initially. As the plume continues to rise, the buoyancy begins to dominate and the vertical velocity rapidly diminishes. The plume rise (which may be substantial as shown in figure 3.12) is thus considered in terms of either momentum or buoyancy or both. Irrespective of the relative importance of these it is found that the effluent often rises to a maximum height at which it then levels off as it is blown downwind (see figure 3.13). The height of the rise, ΔH, is the difference between the final height of the plume above the ground, H, and the height of the chimney, h. This *equilibrium height* is the level at which a source of neutral buoyancy released from rest would produce the identical effect (at ground level downstream) to the actual chimney. This gives rise to the idea of an *effective chimney height* and the replacement of the real source by an *imaginary source* at that height (figure 3.13).

As a plume rises it has a vertical velocity, either directly from its initial momentum, or indirectly from its buoyant rise. At the edge of the plume there is thus a difference of velocity between the plume and the atmosphere. This *velocity shear* produces turbulent eddy motions both in the plume and the atmosphere and these eddy motions mix together the plume material and parcels of air. Plume material is thus transported away from the plume centre and atmospheric parcels are taken within the plume boundary. The result is a more diffuse plume with a larger radius. This process is known as *entrainment*. At this stage turbulence generated by the plume's own motion is more important than any turbulence due to the wind velocity. When the plume's upward rush is retarded and it begins to be blown passively downwind, mixing is continued by these atmospheric eddies and the plume passes into a second and later into a third phase (these last two phases are determined by atmospheric eddies of different scales). The third phase mixing has been shown to be identical with the turbulent mixing assumed in the Gaussian plume model for non-buoyant effluents. When the plume axis is (quasi-) horizontal, the plume is referred to as being bent-over.

Figure 3.12 Under calm meteorological conditions a buoyant plume can rise to great heights: (*left*) downtown Boston (Massachusetts, USA) and (*right*) near Monterey (California, USA).

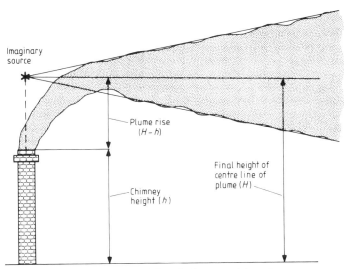

Figure 3.13 Equilibrium height of plume and imaginary source.

PLUME RISE CALCULATIONS

The importance of accurate calculation of the effective height, H, is immediately apparent from equation 3.6 where the concentration at ground level is proportional to $\exp(-\text{const} \times H^2/\sigma^2)$. It can also be shown that the maximum ground level concentration (based on ideas expressed above) is approximately proportional to H^{-2} (a fact first recognised by Bosanquet and Pearson (1936)), although recent measurements indicate that at large heights $C_{max} \sim H^{-2.3}$, and under fumigation conditions $C_{max} \sim H^{-1.5}$. An alternative source suggests that for long period averages $C_{max} \sim H^{-3}$.

Plume rise formulae

Most of the empirical and semi-empirical formulae that exist for plume rise assume that the equivalent stack height is attained whilst the plume is in the first phase, in which case the rise can be determined solely in terms of the initial momentum, initial buoyancy and (in some cases) the windspeed and/or the atmospheric stratification.

(*a*) Neutrally stable case. For momentum dominated plumes, the plume rise, ΔH, is related to the distance downstream, x, by the one-third law, e.g.

$$\frac{\Delta H}{D} = 1.89 \left(\frac{A}{1 + 3/A}\right)^{2/3} \left(\frac{x}{D}\right)^{1/3} \tag{3.13}$$

Briggs (1971) where

$$A = w_0/U. \tag{3.14}$$

For buoyancy dominated effluents, the two-thirds law is found to give a better approximation:

$$z = 1.6(B_0^{1/3}/U)x^{2/3} \tag{3.15}$$

where B_0 is the source buoyancy flux; although this may be less appropriate for smaller, industrial sources (Rittmann 1982).

(b) Stable case. In a stable atmosphere, rise is quickly terminated. There are several formulae available, e.g.

$$\Delta H = 2.3(B_0/UN^2)^{1/3} \tag{3.16}$$

(Perkins 1974). Under calm conditions Briggs (1975) gives

$$\Delta H = 5.0B_0^{1/4}N^{-3/4} \tag{3.17}$$

where N^2 (the Brunt-Väisälä frequency) is given by

$$N^2 = \frac{g}{T}\left(\frac{dT}{dz} + \Gamma_d\right). \tag{3.18}$$

(c) Unstable case. Although theoretically an equilibrium level is never attained for an effluent rising into an unstable atmosphere, some approximate equations exist, e.g.

$$\Delta H = 150B_0/U^3 \tag{3.19}$$

(Smith 1968). This equation may only be appropriate in cases where turbulence is generated by large buildings. Other authors utilise this equation with different coefficients. Alternatively, more recent observations and analyses (see Anfossi 1982) suggest that the two-thirds law (equation 3.15) is also applicable in determining effective plume rise in unstable conditions. In this case the numerical coefficient is lower (about 1.3) to allow for the more rapid dispersion resulting from the greater turbulence levels. It should be noted that these equations (for the neutral and stable cases) assume there is no plume break up due to the existence of large scale eddies (created by, for example, irregular topography, convection) near to the stack.

Typical stable, neutral and unstable plume paths are shown in figure 3.14 in which it is seen that for stable conditions the plume reaches a maximum height (as above) and then sinks back to oscillate about a slightly lower level. This small percentage difference (between equilibrium and maximum or ceiling height) is included in some calculations but the correction is at present less than the error between the models.

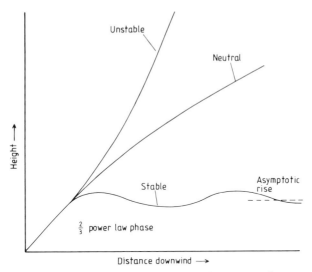

Figure 3.14 Typical three-phase behaviour of buoyant plumes under neutral, unstable and stable conditions. Redrawn from Slawson and Csanady (1971) with the permission of the Cambridge University Press.

Example

To illustrate the difference between these formulae, the results of a numerical calculation are shown in table 3.4 (for a neutral stability case) for an emission of 67.2×10^5 W, initial temperature 380 K, being emitted from a stack 3 m in diameter at a velocity of 10 m s^{-1}. The air temperature is assumed to be 280 K and the windspeed 4.47 m s^{-1}. There seems to be at present little agreement between plume rise formulae. More recent developments in terms of numerical models which are able to predict plume rise through an atmosphere of varying stability (including the important case of inversions) are discussed in later sections.

Table 3.4 Plume rise calculated from four formulae.

Formula	Plume rise predicted (m)
Oak Ridge	39.6
Carson and Moses	53.7
Bryant Davidson	12.0
CONCAWE†	72.1

†Conservation of Clean Air and Water, Western Europe.

Inversion penetration

The height to which a plume will rise is strongly dependent upon the meteorology; especially the vertical temperature profile and the gustiness of the wind. Few of the formulae discussed above are capable of taking into account (in anything other than a very simplistic way) the atmospheric stratification. The presence of an inversion at or just above the stack top can have considerable effect on suppressing the plume rise (see figure 3.15). It is inversion conditions that have the most important effect on ground level concentrations. A necessary prerequisite is the understanding of their formation (see Chapter 2).

Figure 3.15 Regional scale trapping of pollution from many sources (effectively an areal source) can lead to episodic conditions if adverse meteorological conditions persist.

The prediction of the penetration of plumes into or through an inversion layer is not yet well developed. If the inversion layer is relatively thin, then there is a possibility that the plume may 'punch through' the inversion. If it does so it will be trapped above the stable layer and pollutants will be unable to reach the ground until the inversion is broken. If there is no wind, the plume rises vertically and will penetrate the inversion if its temperature excess when it reaches a height H_i (the height of the base of the inversion above the source) is greater than the temperature increase across the inversion layer ΔS_i. This is admirably exemplified for the power station plume in figure 3.15 which is much more buoyant (and emitted from a greater height) than emissions from surrounding industry which remain trapped by the inversion layer.

Negatively buoyant plumes

Practically all the work done on plume rise and plume diffusion assumes that both momentum and buoyancy are positive quantities (or at worst zero). Of course, when a plume rises under stable conditions it overshoots its equilibrium level and continues until it loses its momentum. At its maximum height it is negatively buoyant and begins to descend and oscillate about its final equilibrium height (the level at which it has zero buoyancy). Plumes emitted vertically upwards from large chimney stacks are unlikely to have negative momentum. However it is quite feasible for them to be negatively buoyant (e.g. the accidental release of dense gas at Seveso in 1976). The buoyancy force acting on a plume is usually calculated by consideration of the temperature excess. This is only equivalent to a density deficiency (and thus buoyant effect) if the density of the plume and of the air at the same temperature are identical (i.e. it is assumed that the molecular weight of the plume is equal to the molecular weight of the air). For many plumes containing more than 90 % air in the waste gases this is probably not a bad approximation.

There are however many cases in which the effluent is either cooler than the air at the chimney top height or denser than air. In these cases the equilibrium level may be beneath the emission level. The same arguments used to discuss a buoyant plume above its equilibrium height but below its ceiling height (i.e. positive momentum, negative buoyancy) can be applied to this situation. Subject to the conditions of inversion penetration described above, the plume will rise until all momentum is lost and then descend to its equilibrium level. However, under unstable conditions it is possible that an equilibrium level will be attained during the ascending phase. In general the plume is likely to be hotter but denser (i.e. a higher molecular weight) than its environment. The resulting buoyancy may be positive or negative. This can be easily taken into account by use of numerical models (see discussion below).

Numerical models

Theoretically derived mathematical models have been developed over the last twenty years. They are almost all based on hydrodynamical conservation equations together with a necessary 'closure' assumption † and attempt to describe the steady-in-the-mean plume. The classic paper by Morton *et al* (1956) forms the basis for many of the models of plume rise. There are five conservation equations (mass, two components of momentum, buoyancy, energy): four of these are necessary and one redundant. Most models are also only strictly applicable to weakly buoyant plumes in which the Boussinesq‡ approximation can be used to simplify the analysis, although

---

†An assumption relating two or more variables needed to augment the number of equations for there to be a mathematical solution (namely number of equations equals number of variables).

‡The Boussinesq approximation states that all density differences can be neglected unless they occur in the buoyancy term (i.e. occur multiplied by g).

one recent model eliminates this restriction. To allow for pressure changes, densities and temperatures are replaced by potential densities and temperatures although this refinement is seldom stressed.

The models often use an entrainment assumption to describe the rate at which material is taken into the plume, thus diluting and spreading it. Different rates are envisaged to describe the rising phase in which self-induced turbulence is created by the velocity shear as the plume initially rises rapidly through the relatively quiescent atmosphere, followed by mixing by atmospheric eddies as the plume is bent over and the final phase mixing equivalent to the Gaussian plume model is approached. The results of these models show (for neutral conditions) a first phase rise in which $z \propto x^{2/3}$; a result borne out by empirical models and observations.

Urban plumes

A further complication that may arise in either an unusual topographic situation or in an urban area is an anomalous vertical temperature profile; e.g. when a low level inversion occurs at differing heights within the urban area. Rising air within the city forces incoming, rural air (in which the temperature structure is often that of a low level or ground based inversion, especially at certain times of the day) to rise over the city 'dome'. Air within the dome is trapped and circulates within the confines of the urban 'heat island' (see figure 2.23 and Chapter 2). A cross section through a city reveals an inversion height, initially small, which rises at the city centre, usually above stack heights and provides typical inversion conditions as described above. Effluents are easily trapped at or below the inversion height. An increase in chimney height will of course increase the effective stack height, but not proportionately for emissions which penetrate into the inversion layer. Table 3.5 shows that an increase of 90 m in chimney height results in an increase in equilibrium height for the plume of only 70 m.

Table 3.5 Simulated plume rise heights for an effluent emitted in the urban area of Leeds (UK) at a temperature of 300 K and a velocity of $0.5 \mathrm{~m~s}^{-1}$.

Chimney height (heights above 10 m in brackets)	Inversion height (above stack top)	Height of rise (at 5 km downstream)	Total height above ground	Additional height (cf 10 m high chimney)
10	44	44.8	54.8	—
25(15)	29	43.7	68.7	13.9
50(40)	4	35.9	85.9	31.1
100(90)	−46	25.0	125.0	70.2

When pollutants are trapped, they give rise to an increase in the concentration in the 'pollution dome' in which they can be assumed to be well mixed (homogeneous). Both *box* and *slug* models can be used to calculate concentrations within the urban environment as a function of emission rates and removal by the wind (advective losses) (see, for example, Benarie 1980).

Radiation effects and increased windspeeds can destroy this heat dome and release pollutants from the confines of the city. Then the city has a plume. City plumes have been observed, as also have continental plumes. For example dust and insect swarms typical of Africa have been observed in a plume extending across the Atlantic towards the Americas.

The many factors involved in plume trajectory and plume rise calculations (e.g. meteorology, chimney height, efflux characteristics) are found to interact strongly and it is not always possible to isolate the major factors determining plume paths and ground level concentrations. Once the plume path is known, however, the pollutant concentrations are easily calculated by assuming a Gaussian distribution. Figure 3.16 shows the ground level concentrations calculated for an emission of 0.1 $kg\,s^{-1}$ of sulphur dioxide over flat terrain and into a neutrally stratified atmosphere. The maximum value occurs at about 350 metres from the source which is about eleven times the stack height in this example. (Empirical results suggest a maximum occurring between ten and fifteen stack heights downstream.)

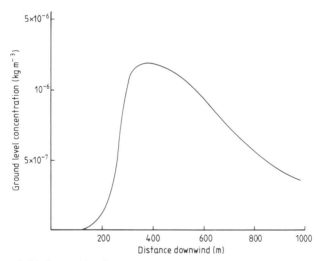

Figure 3.16 Ground level concentrations as a function of distance downwind calculated using a numerical model of an emission of 0.1 $kg\,s^{-1}$ of SO_2 over flat terrain and with a neutral atmosphere.

Complex terrain

Larger obstacles are known to disturb the flow pattern of the air moving over them. A quantitative assessment of the hydrodynamics of this effect is outside the scope of this discourse. Qualitatively, the streamlines over the hills tend to be compressed, resulting in an actual plume height (above ground level) that is reduced. In neutral and unstable conditions, this is likely to be the dominant feature of the plume path in which the ground level concentrations are increased, but the plume does not impinge on the hillside. Under stable conditions, higher pollutant concentrations may be observed. When ventilation is sufficient (i.e. moderate windspeeds), the problems of plume impingement may be several. High concentrations will be observed (possibly as high as plume centreline values) but for relatively short periods, although these may be sufficient to cause damage to plants and distress to animals and humans. In the case of the inversion base below the stack top, then pollution is trapped in the stable layer. As the plume approaches a hill, the base of the inversion layer tends to rise as the streamlines begin to pass over the hill. The inertia of the plume may well cause it to carry on in a straight direction so that it effectively penetrates into the mixing layer. Fumigation follows and high levels of concentration are observed on the hillside. In the case of (semi-) isolated hills it is possible that the plume will not impinge but flow laterally around the hill (under stable conditions).

The case described above is an extreme example since irregular terrain also introduces additional turbulence into the air flow which dilutes the plume more rapidly. Clear air turbulence (CAT) is observed to occur with greater frequency in mountainous regions than over flat terrain—especially if the flow is well stratified. This once again dilutes the plume rapidly, although it may dominate sufficiently to transport vertically discrete 'lumps' and create a meandering or looping plume.

The second constraint on pollutant dispersal in a valley has perhaps had more large scale repercussions over the years. If the area is sheltered topographically, then the advent of a slow moving anticyclone (see Chapter 2) often results in complete atmospheric stagnation—with virtually no ventilation (i.e. zero wind), clear skies and inversion conditions. The resulting episode may have dire consequences. Well known examples of this phenomenon occurred in the Meuse Valley in Belgium in 1930 and in Donora, Pennsylvania in 1948. In such circumstances pollutants are trapped by the inversions (as discussed previously) and concentrations start to increase since there is no dispersive mechanism. Initially pollution is trapped at the level of the inversion (see figure 3.15) but will eventually diffuse slowly throughout the whole depth of the lower layer.

REMOVAL OF POLLUTANTS

Dry deposition

Particles in a plume are subject to gravity and tend to fall at a terminal velocity, V_0, determined by Stokes's law (see Chapter 1). This velocity is an increasing function of size such that only relatively large particles are deposited—gas molecules remain in suspension longer and reach the ground by the diffusive processes outlined above, when they too may suffer dry deposition. The ground level concentration due to large particles is thus enchanced near the stack. They are found to settle at a distance $x_g = UH/V_0$ from the point of emission. Deposition can be modelled by use of a *tilted plume model* (see figure 3.17) in which the value of H is replaced by $H - xV_0/U$. The fact that material has been lost from the plume can be included in the model by depletion of the source at a rate downstream corresponding to the deposition rate, or can be represented by using a surface-depletion model.

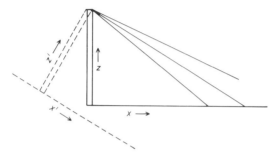

Figure 3.17 Tilted plume model. Diffusion from a real chimney is assumed to be identical to that from a tilted chimney (broken line). Concentrations are calculated as before in coordinate system (x',z') in which the plume centreline is parallel to the ground (broken line). The axes are then transformed to give concentrations in (x,z) coordinates in which the plume centreline is at an angle to the ground.

Washout and rainout

In addition to dry deposition by gravity and diffusion by turbulence, plume material (both gases and particulates) may be removed from the atmosphere by washout and rainout. When rain droplets fall through the plume they collide with molecules in the plume. These particles are often captured (with less than 100 % efficiency), removed from the plume and brought to the ground (*washout*). Gases (e.g. SO_2) can also be removed in this way. Some gases may react with the rain and form an acid solution—carbon dioxide is removed from the atmosphere by falling rain which becomes a weak solution of carbonic acid (H_2CO_3). Sulphur dioxide, once oxidised to SO_3, may react similarly thus giving rise to the problem of acid rain (H_2SO_4). This is often enhanced by the presence of HNO_3 formed as a result of the large emissions of NO_x into the atmosphere (see also Chapter 1). Complaints about the effects of such acid rain in the neighbourhood of a gold smelting industry in Western Australia appeared justified since clothing dissolved rapidly in the rain and even the chimney itself was regularly replaced.

Calculation of the amounts of pollutants that can be removed from the plume by washout are complicated by the need for information on the spectrum of rain droplet sizes. Hales *et al* (1973) derive a simple linearised model which uses equation (3.6) together with an integral equation for the raindrop sizes and conclude that the average concentration of pollutants in rain is given by

$$C_{av} = F_0/J \qquad (3.20)$$

where J is the rainfall rate in cm s^{-1} and F_0, the pollutant flux at ground level, is calculable in terms of their model.

An alternative removal mechanism for particulates is *rainout* in which particles act as nuclei on to which water vapour can condense when the air is saturated. Once condensed the raindrops may grow rapidly by further condensation or by collisions until they are large enough to settle out by gravity.

Wet plumes
Any plume which contains some water vapour is liable to condense and become visible. Condensation occurs when the mass of air becomes saturated. This becomes increasingly likely as the plume rises and cools since the maximum amount of water held in a plume in the form of vapour decreases with decreasing temperature. As water vapour condenses it releases energy in the form of latent heat. This extra energy makes the plume more buoyant and enhances the rise. Until the plume is saturated, however, no condensation occurs and the plume rise is identical to that of a *dry plume*. Once condensation is initiated the effluent is known as a *wet plume* or a *saturated plume*. (Most plumes containing water vapour reach this point.) Although a wet plume rises to a greater height than a dry plume, this is only true if all other efflux parameters are identical. In many plumes water is present because the waste gases have been scrubbed to remove pollutants (see Chapter 8). This process cools the gas and thus the emergent wet plume is usually much cooler than it would otherwise be. It is thus less buoyant and its net rise may be less than it previously was.

SUMMARY

The dilution of an effluent depends upon three factors: the spreading (dispersion) due to the wind, the height to which the effluent rises, and chemical and photochemical reactions which may produce secondary pollutants and change the concentration of other species. Diffusion can be calculated from a basic Gaussian plume model over flat terrain. The effects of complex terrain are still under investigation. The shape of the plume depends upon the stability (see Chapter 2) as does the extent of the plume rise.

The rise of a gaseous effluent is determined by its momentum and buoyancy and also by the meteorological conditions (wind and vertical temperature profile). If the atmosphere is stable (or possesses an elevated stable or inversion layer) the plume rise will be restricted. In unstable conditions the plume's buoyancy increases as it rises and (theoretically) the plume may never reach an equilibrium level. Mixing of the plume and ambient air occurs because of self-induced turbulence which entrains air into the plume and by turbulent eddies either within the wind or generated by the underlying topography.

Absolute prediction of plume trajectories is still less than certain. Currently, empirical formulae are used for engineering design, although the calculated values of plume rise may differ by a factor of two or three. Numerical (computer) simulation is difficult and is not yet in total agreement with observations or simple formulae. However these models are able to take into account at least some of the effects of the underlying topography.

4
Monitoring of Pollutants

The assessment of the degree of harm or nuisance caused by an air pollutant, either whilst it is still airborne or after removal and deposition on objects on the ground, introduces the problem of quantification. It is frequently possible to devise a laboratory experiment to deduce, for instance, the effects of pollutant X at a known and constant concentration for a measured duration on a specific crop. Transferring that knowledge to a real situation is not easily accomplished since the inherent variability of pollutants (especially as a result of turbulent meteorological characteristics) strictly requires continuous four-dimensional monitoring, usually for periods up to several years. Such unbroken surveillance in three spatial directions and time is impractical except for very small-scale field observations.

The fluid's continuum properties necessitate a *sampling programme* and the subject of devising such a system is extensive and very important (see, for example, Munn 1981). The design must reflect the aim of the study. For instance, the Warren Spring co-ordinated National Air Pollution Survey in the UK measures mean daily values of smoke and SO_2. Although there are 365 items of data per station (well over 1000 in the UK, reduced to 150 in April 1982) per year, the usefulness of this data base is in examining trends over a period of a few years and says little about either temporal or (interstational) spatial variability on a small scale.

Indeed the biggest problem in both collection and interpretation of data is the temporally and spatially variable nature of most pollutants. The sampling scheme must bear these factors in mind, together with the specific objectives of the survey (namely local 'hot spots', long term regional trends etc). At a given observation site, sampling may be undertaken continuously, by instantaneous sampling at each fixed interval (e.g. one sample every hour) or by time weighted averaged (TWA) sampling procedures (e.g. UK National Survey) which effectively integrates the total pollutant received over a selected time period (e.g. mean daily amounts). Some form of time series analysis may then be appropriate. Many survey questions are, however, related to spatial variability (usually in two horizontal dimensions and less frequently in the vertical direction). Here the question centres on the representativeness of the site for the local region. The World Meteorological Organisation (WMO) defines three types of monitoring surveys: baseline (in remote areas to ascertain a global background level), regional (rural) and stations dedicated to a specific purpose. This latter group will often be sited in high concentration urban areas where positioning is difficult due to local inhomogeneities and microscale meteorological influences. Much work is still being undertaken to try to evaluate the representativeness of single sites (for their local area) and the usefulness of observing networks for evaluating regional pollutant concentrations (e.g. Henderson-Sellers 1980, Munn 1981). Work has been undertaken to assess, statistically, the number and siting of stations in order to achieve the prescribed objective—to obtain a true monthly mean within specified limits of accuracy. Another frequent problem is the sitting of additional stations within an existing network. Buell (1975) suggests that this can be accomplished by identifying the location which has the largest interpolation error of the existing network.

It must also be stressed that many instruments need calibrating prior to use. This is especially important for continuous gas analysers. This may be in the form of a thorough or primary calibration, taking several hours, or a routine two-point check (secondary calibration) which is performed more frequently, usually every few months. Here the instrument response is evaluated for samples of (*a*) 'clean' air and (*b*) a high concentration of a single (known) pollutant to check the response at the two detection limits. Furthermore the response time of the instrument must be matched with the time resolution desired for the sampling programme. Slow response monitors cannot be used to monitor continuously without losing resolution in a rapidly varying pollutant concentration. The accuracy of different instruments can vary by orders of magnitude, as can their lowest level of response (i.e. ability to measure low concentrations). A rapid, cheap and perhaps relatively inaccurate survey may be sufficient to delineate a small area as being in particular 'danger'. This survey would then indicate a more substantial monitoring programme without the necessity and cost of instrumenting the total area.

Recently much work has been undertaken into discovering the fate of pollutants transported over long distances. The acid rain problems of Scandinavia and Canada in particular have stimulated research programmes such as STATE (Sulfur Transport and Transformation in the Environment) established by the USEPA. The development of the tar sands of Alberta has been preceded by a five year monitoring programme for meteorological parameters and background pollution and there will continue to be extensive monitoring on a variety of time and space scales (for sulphur dioxide, dust etc) as the project develops.

Pollutant effects may be governed by long term average exposure; yet for certain pollutants exposure to a high, but short-lived, peak may have disastrous consequences. Before discussion of the harmful effects (see Chapter 5) it is therefore necessary to discover how, or indeed if, different pollutants can be measured *in situ*, at the point of emission (or in the stack just before emission) or in confined environments. The present chapter discusses the basic concepts in sampling and measuring, either *in situ* or by collection and subsequent laboratory analysis. The subject of data interpretation and analysis will be discussed briefly and some available pollution data re-examined.

GRAB SAMPLING

Collection of samples for laboratory analysis requires a container and a method of removing the sample from its environment. The simplest method is to use an evacuated bag or bottle so that when the seal is opened, air enters as a result of the vacuum. Alternatively a pump can be used to draw air into the bag. For pollutants of low concentration it is more useful to pass the air across an adsorbing medium so that the pollutant is concentrated sufficiently for detection by standard laboratory methods. Once in the laboratory this must be removed from the collection medium and passed on to the measuring element (e.g. a gas–liquid chromatograph). This is often accomplished using thermal desorption techniques, in which an electric current volatilises the sample which can then be swept off the adsorber by using a neutral gas stream.

THE SAMPLING TRAIN

Figure 4.1 shows the basic components of a sampling train. The order of subsystems is important here with the air mover *pulling* air through the system. All pollutant monitors can be described in terms of these basic components, however simple the monitor appears to be. Each section in figure 4.1 will be discussed in detail below. A sample is abstracted into the monitor at the inlet. This may be simply an orifice or a carefully designed receptor hood. The second stage is the quantitative (or sometimes qualitative) assessment for the (usually specific) pollutant under investigation. Not until after this stage is a measure made of the total volume of air from which the pollutant is trapped by the measuring device itself (this may be a gas meter or calibrated orifice). The air mover is usually a small pump and sucks air through at a constant or predetermined rate. Pumps should never push air through a sampling train since this is likely to lead to high pressure conditions if a blockage occurs. The outlet for the system should not, of course, be positioned upwind of the inlet.

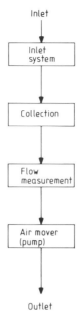

Figure 4.1 Schematic diagram of a sampling train for air pollution monitoring.

At each stage, detailed considerations are needed to select the most appropriate piece of apparatus; the potential influence of chemical, dynamical and meteorological factors, for instance, will often need to be accurately assessed, bearing in mind the reliability, accuracy and precision available and /or required by the programme's objective.

In the sections that follow a brief account will be given of each of these factors, although it is impossible (within a single chapter) to list or discuss every monitoring approach to all pollutants of even widespread interest.

Sampling inlet
The inlet is frequently an inverted funnel. This will collect a gaseous sample together with any suspended particles below the size range which will gravitate out rapidly. At the same time, infiltration of precipitation is largely avoided. The efficacy of the sampling will be determined to some degree by the funnel size as well as its orientation with respect to the wind direction and speed (see later discussion on isokinetic sampling). Liu and Pui (1981) have reviewed some of the sampling inlet types for both total suspended particulates (TSP) and inhalable particulate matter (IPM).

Measuring element
This can perhaps be regarded as the cornerstone of a sampling train. Any assessment may be made on a continuous or an intermittent basis; the latter

by means of a clock mechanism which activates the monitor at regular intervals. Quantification may be performed *in situ*, or it may be necessary to collect the pollutant and remove it from the train in order to evaluate the pollutant concentration (e.g. by analytical chemical techniques). It is convenient to divide the many hundreds of samplers into two groups: those for measuring gases and those for measuring particulates. Indeed the methods are closely akin to those used in air pollution control (see Chapter 8). Some commonly encountered examples can then be described in the light of these basic ideas.

Gas monitoring. Gases cannot be removed by any gravitational techniques. Use must be made of either their chemical properties or their diffusive nature. Many gases are collected by absorption or adsorption (see Chapter 8); or their *effects* can be monitored directly, e.g. in the retardation of plant growth.

Figure 4.2 The lichen *Lecanora muralis* is relatively tolerant of SO_2 levels and can thus be used as an indicator species in urban areas.

The simplest monitoring devices are static samplers which possess no moving parts. They rely on a chemical or biological interaction between the pollutant and the sampling medium, although they may also be affected by climatological trends as well as pollutant trends. Lead dioxide candles (now virtually obsolete) were formerly used to monitor SO_2, which changed the lead coating on the candle to lead sulphate. Mosses and lichens can be used in a bioassay to infer the ambient concentrations. The use of the lichen *Lecanora muralis* (figure 4.2) in the UK has indicated how SO_2 concentrations in urban areas have changed over several years (see figure 4.3). Decreasing SO_2 levels not only permit the invasion of *L. muralis* but further from urban areas, the comparatively cleaner air will allow more luxuriant forms to grow (figure 4.4). However the lichen scale of Hawksworth and Rose (1970) may not be applicable in other countries (van Haluwyn and Lerond 1983) where the 'stresses' (see table 5.2) may be different.

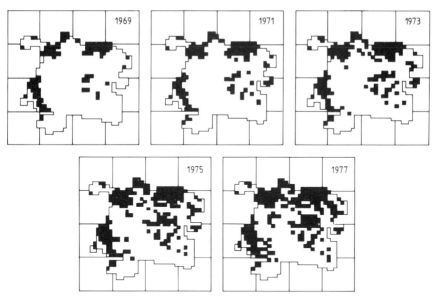

Figure 4.3 Digitised field data showing the areas which *Lecanora muralis* has reinvaded (shaded). Data are for the West Yorkshire (UK) conurbation. After Henderson-Sellers and Seaward (1979) with permission.

Figure 4.4 More luxuriant lichen forms (cf figure 4.2) flourish in areas where the atmosphere is cleaner (i.e. less SO_2). Left: *Hypogymnia physodes*. Right: *Parmelia saxatilis*.

The effects of H₂S are noticeable in the discolouration of paint and lead acetate paper. Concentrations cannot be measured accurately with this method, but some overall indication may be gained.

Odours can be detected by the nose and by fabric discolouration (e.g. as a result of exposure to ozone).

Chemical methods are exemplified by the standard wet chemistry techniques for SO_2 evaluation. The quantification (titration is often used here) must be matched with the collection procedure. The pararosaniline method is employed by the USEPA whereas the hydrogen peroxide method has been adopted for use in the Warren Spring National Survey in the UK. In the latter case the 'bubbler' (see figure 4.5) contains a hydrogen peroxide solution. Almost all SO_2 in the air flow is removed by the reaction

$$H_2O_2 + SO_2 \longrightarrow H_2SO_4.$$

However there is some interference from other soluble gases and particulates and in general a pre-filter is used to remove particles such as smoke. In the pararosaniline (or West-Gaeke) method, the SO_2 is absorbed in a dilute aqueous solution of sodium tetrachloromercurate. Addition of formaldehyde and bleached pararosaniline creates a red-purple colour, the intensity of which is proportional to the SO_2 concentration. (This value can be reduced markedly if ozone or nitrogen dioxide is also present.) These techniques may be used on different time scales.

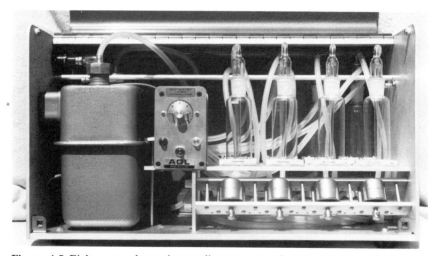

Figure 4.5 Eight-port volumetric sampling apparatus for SO_2 sampling. Photograph courtesy AGL Engineering Development Ltd, Hitchen, Herts, UK.

The basic (e.g. hydrogen peroxide) bubbler does not claim any short term accuracy and is most useful in determining daily averages. In this mode of operation, it is often desirable to use an eight-port—a range of eight

Figure 4.6 Miniature SO$_2$ sampler. Photograph courtesy C F Casella and Co., London.

identical bubblers with a switching device so that one week's samples may be taken before analysis (the eighth bubbler is for emergency use). The advantage is in saved man hours; yet it is possible that some chemical deterioration may occur during the period of storage of up to one week. The relatively low cost of the bubblers permits a wide network to be established rapidly across a country. Miniature, portable versions are also available (figure 4.6) in which sampling can be undertaken over a 24 hour period with observations from one to eight times per hour. (An additional recording instrument must be attached for hard copy provision.) However to measure, on a continuing basis, short period (of the order of a minute) pollutant peaks, which may be an order of magnitude greater than the mean daily value, continuous monitors have been developed. These operate on a coulimetric, conductimetric or flame photometric method. The latter method requires a hydrogen supply and hence needs more stringent safety precautions for its use. Results are in good agreement with 24 hour bubbler values. Resolution over shorter periods is good, but capital and running costs are higher. Figure 4.7 shows a trace from a continuous SO_2 monitor, showing the wide variability inherent in the observations. The (possibly important) loss of information by use of longer sampling periods is shown in figure 4.8 where the information in figure 4.7 is integrated over several time periods. Health effects may be related to long term average concentrations and/or short period peak exposures (see Chapter 5).

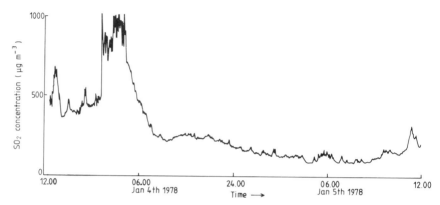

Figure 4.7 Continuous trace for ambient SO_2 concentrations, 4/5 January 1978, Salford, UK.

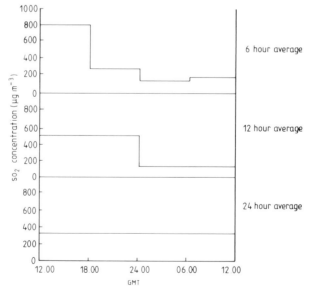

Figure 4.8 SO_2 values from figure 4.7 integrated over three different time periods.

Gas–liquid chromatography (see figure 4.9) is based on the principle that passing a mixture of gases along a packed column (containing a stationary phase coated on packing material, often consisting largely of diatomaceous earth) will result in a differentiation since the stationary phase slows down the passage of each gas in a characteristic way. The time taken for the gas to pass through the column permits identification—although linking a gas–liquid chromatograph to a mass spectrometer permits even more rapid identification.

Lasers can be used not only as rangefinders but also to determine the chemical constituents. The laser-based LIDAR and the acoustic SODAR instruments are both based on the collection of backscattered energy and are currently being used in experimental studies. A ground based microprocessor ultraviolet television sensing system has recently been introduced (VISIPLUME) to give real-time analysis of sulphur dioxide content of plumes from ground based observations.

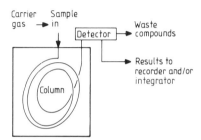

Figure 4.9 Diagram and photograph of a gas–liquid chromatograph.

Radioactivity can be monitored by Geiger–Müller detectors, and ionised gases emitted from some stacks are now being used to trace the path of atmospheric pollutants.

Particulates. Particulates can also be monitored rapidly, simply and cheaply using static samplers. The ability of particles to be removed from the atmosphere by gravity permits collection plates to be used with effect. A plate, covered with gel, will prevent wind disturbance to collected material, but precipitation can thwart this sampling procedure—even under ideal circumstances it is not feasible to assess the size ranges, nor to use them for anything more precise than a rough survey. The standard deposit gauge (see figure 4.10) is considered to be obsolescent by many, despite its long acceptance (perhaps because of the lack of anything better). The area presented to the atmosphere is about 0.312 m^2 and collected dust is retained in a bottle suspended beneath the funnel. Once again precipitation causes difficulties as does the possibility of chemical reactions since the measurements are only undertaken once a month. The contents of the bottle can be analysed for pH, total undissolved matter, 'tarry matter', ash, combustible particles (excluding tar), total dissolved matter, calcium, chloride, sulphate and total solids.

Figure 4.10 British Standard deposit gauge in operation on a rooftop.

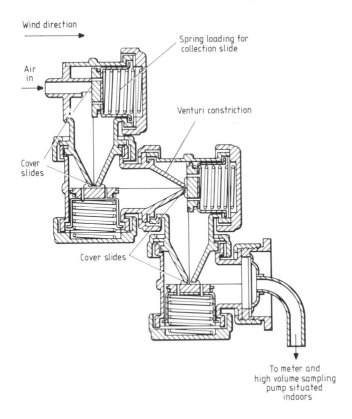

Figure 4.11 Schematic diagram and photograph of a four-stage cascade impactor for identification of the size of collected particulates.

The inertial properties of particulates are also used to good effect in devices which rely on impingement, as the air flow is forced to turn through a right-angled bend. In one type of cascade impactor (figure 4.11) this occurs four times. As the size of the particle removed depends on the air velocity, the air approaching each bend is channelled through a venturi constriction, thus permitting the velocity of approach to each of the four collection plates to be regulated individually. Collected particles can then be analysed microscopically or gravimetrically or removed from the plate for further chemical tests.

Filters are regularly used to remove particles from a gas stream. Smoke measurements are taken in the UK as part of the National Air Pollution Survey using a clamp fitted with a Whatman No. 1 filter paper as part of a 'low volume' sampler. In the 1950s, when it began, this 'smoke' survey was designed to measure total suspended particles (TSP) of which the largest percentage were smoke (as designated in Chapter 1, namely carbonaceous particles in the approximate size range $0.01-1.0$ μm). This *Standard Smoke* value was determined from calibration curves of reflectivity of TSP samples collected in London. Today Standard Smoke is still measured, but because its composition is drastically different from that of 25 years ago, it is doubtful (Ball and Hume 1977, Bailey and Clayton 1982) whether the value calculated is still a reasonable measure of TSP. (The USEPA has adopted the high volume sampler as its recommended method. This has been shown (Pashel and Egner 1981) to give values approximately 105 μg m^{-3} higher than the British smoke sampler.) Particulates are

Figure 4.12 Photoelectric 'smoke stain' reader measures density of blackness of 'smoke' stain.

removed not only by a sieving effect, but also by inertial effects, diffusion and impaction of smaller particles (see Chapter 8 for a fuller discussion). Used in conjunction with an SO_2 bubbler, the filter must be changed daily or automated. Measurement is performed by reflectivity assessment (figure 4.12) or gravimetrically. Smoke sampling *per se* is decreasing in importance as ambient concentrations fall. Black smoke emissions from chimneys beyond the statutory limits may still be assessed visually. The Ringelmann chart (figure 4.13), held at the correct (i.e. dependent on the size of the chart being used) distance away, gives a representation of the opacity (blackness) of the smoke, since at a distance the cross-hatched lines of the chart blur together to give an impression of overall grey—the intensity of which depends upon the thickness of the cross-hatching. Some legal requirements are assessed by such visual matching to give a Ringelmann smoke density.

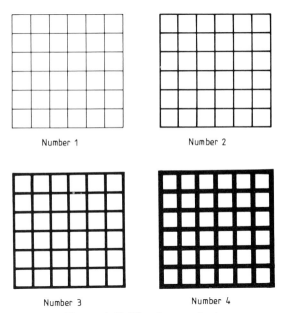

Figure 4.13 Ringelmann chart.

Of increasing interest are particulates less than about 10 μm† in size, since these can be inhaled (see Chapter 5). Removal of such particles from an air flow may often be undertaken by thermal or electrostatic precipitation. The latter is used on a large scale in control devices (see Chapter 8) and for monitoring. Figure 4.14 shows a respirable mass monitor which first removes large particles by an inertial impactor (the cut-off size can be selected) and then precipitates smaller particles on to a piezocrystal.

†The USEPA (Miller *et al* 1979) define IPM (inhalable particulate matter) as comprising particles with aerodynamic diameters $\leqslant 15$ μm.

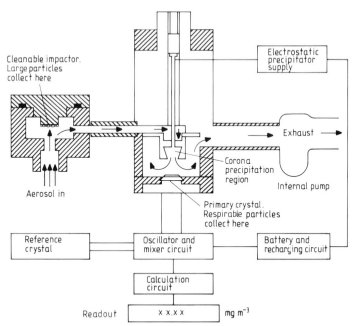

Figure 4.14 TSI Respirable Aerosol Mass Monitor Model 3500. Photograph courtesy BIRAL(TSI), Portishead, Bristol.

Changes in resonant frequency of this crystal are monitored and the net change over a measuring period of two minutes (24 seconds for heavily dust-laden atmospheres) is then converted to an average concentration in mg m^{-3}. No differentiation of size can be made and hence the accuracy of the results is not high. Nevertheless it is a useful and quick method for use in a 'walk through' survey to highlight hot-spots.

There are of course a very wide range of instrumental and analytical techniques available in the laboratory for determination of atmospheric pollutants. It is not possible to discuss all these methods, many of which could be considered as basic analytical chemistry (including wet chemistry methods). However there is one instrumental technique worthy of some brief discussion—spectroscopy. A spectrometer (which may use visible or infrared radiation) is an instrument in which the sample (usually in solution) is either injected into a flame which is often acetylene (although this will depend to a large extent upon the element that is being quantified) or put in an oven in which it is atomised. Only one element can be identified at a time, by irradiating the atomised sample with radiation of a specific wavelength (i.e. specific to the element under consideration). Any pollutant present will absorb that radiation. The greater the concentration, the greater the degree of absorption. Measurement of the degree to which the radiation beam has been depleted (usually by means of a photomultiplier and digital display) can be used to quantify the concentration of the pollutant in the sample. An atomic absorption spectrometer is shown in figure 4.15. This instrument is designed to look specifically for heavy metals and is used widely in atmospheric pollution studies for lead and, to a lesser extent, cadmium.

Figure 4.15 Atomic absorption spectrophotometer.

In the case of laboratory analysis, the sample must be collected. There are two basic methods: grab sampling and continuous sampling. Grab sampling is any sampling method which obtains a single sample at a single point in space and time. It thus does not pretend to give anything other than a rough indication of the presence of pollutants. However grab sampling, despite its limitations, is often the only method available (within financial constraints) to obtain the sample. Grab sampling can be accomplished simply by filling a previously evacuated bottle or indeed by spot measurements by instrumentation, which does not require the sample to be brought back to the laboratory (e.g. Dräger tubes). A more sophisticated method is to integrate in time, if not in space. Indeed this is often necessary when ambient concentrations are so low that any straightforward grab sampling method would not provide sufficient amounts of the pollutant material to enable any field or laboratory method to identify it. Passing a continuous stream of air through a container filled with, for instance, activated carbon, will cause the pollutant to be adsorbed on to the collection material, thus effectively concentrating it. Once back in the laboratory this can be desorbed by one of several methods and the pollutant identified.

Metering device

The third section of the train is any device for metering the air flow. Due to the relatively large volumes of gas to be measured, one of the commonest methods is to measure the total volume of air passed through the measurement device during a predetermined time interval. This volume can be measured by, for example, a spirometer or gas meter. Alternatively an evaluation of the volumetric flow rate can be made using, for instance, a venturi meter, limiting orifice or rotameter. A less common alternative is to permit only a specified amount to be drawn through the apparatus by the air mover at each (discrete) operation. Two well known pieces of apparatus will help to illustrate the mode of action of metering devices.

The Warren Spring eight-port includes a basic gas meter. Over the period of operation (here the sampler is changed effectively once a day) the airflow is monitored and the total flow $V_{i+1} - V_i$ (where V_i is the meter reading at the beginning of the ith day of operation) calculated by the difference between the two meter readings. The accuracy is thus the accuracy of the meter itself (although not the overall accuracy of operation of the eight-port). On some days random variability in, for instance, the power supply may result in different values for flow rates over successive 24 hour periods.

Operation of Dräger tubes, on the other hand, requires only two pieces of apparatus. In fact each piece performs two of the functions within a complete sampling train. The tube itself is both inlet and measure; whilst the vacuum bellows act as an air mover and a meter. Here each operation of the hand bellows causes a constant volume (10^{-4} m^3) of air to be pulled through the measuring tube. The metering facility, being inbuilt, is thus

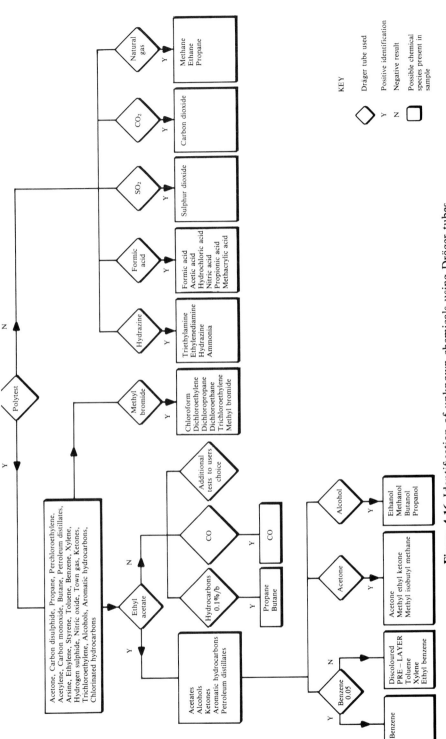

Figure 4.16 Identification of unknown chemicals using Dräger tubes.

more difficult to check for continuing accuracy; although Dräger tube assessment is never intended to provide pollution measurements of the highest accuracy since any errors inherent in the metering operation of the equipment are likely to be of a smaller magnitude than other possible errors. Each tube is specific and prior knowledge (or suspicion) of the presence of that pollutant is required for successful use. However Dräger has recently developed a series of tests to identify pollutants present in a previously unidentified mixture (figure 4.16) either for ambient sampling or industrial hygiene applications (for which purpose they are primarily intended).

The mass monitor's meter too is inbuilt, relying on a constant volume pump together with an accurate timer to maintain constancy throughout a lifetime of two minute or 24 second measurements.

Air mover

The air mover is almost always synonymous with a pump and most of these aim to maintain a constant flow rate with time. The flow rate may be different for different pump requirements. For example, there are monitoring techniques which require low volumes of air (e.g. UK SO_2 bubblers) and those which are essentially high volume samplers (some USEPA recommended techniques). If the pump is considered at all unreliable strict watch must be kept on the volume flow at the metering stage. The constituents in the collected sample are a function of the flow rate. In general, higher rates permit the collection of larger particles, although the relationship is not a simple one and many other factors (e.g. inlet type, see above) are involved. The flow rate will also (partly) determine the capture efficiency of particulates by a filter paper. Research into these variables and their interrelationships in the case of airborne lead sampling is currently being undertaken. Pumps capable of supplying variable flows can also be used in this type of work. In all cases reliability is required, as for many monitoring programmes 24 hour sampling over periods of years is needed—access to a power source is thus often a prerequisite of the sampling site. Battery operated pumps, often as part of a portable instrument (e.g. the portable SO_2 sampler shown in figure 4.7), need frequent recharging and their field life is usually very limited (frequently to less than 24 hours).

STACK SAMPLING

The foregoing discussion has dealt with sampling procedures for atmospheric samples. A second important type of sampling is that of flue gases whilst in the stack or in ducting between the combustion chamber and outlet. In many instances this can be a hazardous experiment and skilled steeplejacks may be required to employ the measuring devices. Concentrations are of course much higher and this may introduce the problem of toxicity of the gas being sampled. In addition to sampling for the pollutant itself it is also necessary to evaluate characteristics of the gas flow. Measurements are thus taken of the gas velocity, temperature and pressure as well as concentrations, windspeed and direction and physical size of the ducting or chimney.

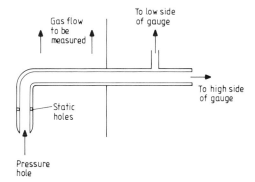

Figure 4.17 Standard Pitot tube.

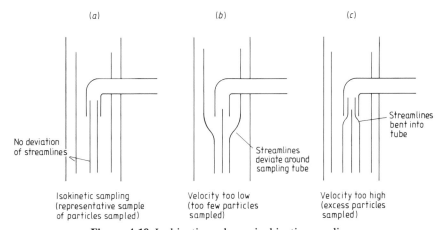

Figure 4.18 Isokinetic and non-isokinetic sampling.

The flow patterns within a duct are complicated by boundary (wall) effects and the fluid dynamics of flow round bends. It is always necessary to find the longest straight piece of piping for the sampling location at which transverses can be undertaken. Pressure distribution is often measured by a form of Pitot tube (figure 4.17) and temperature in terms of both dry and wet bulb temperatures. Samples can be withdrawn for further examination. In this case it is always necessary to emply *isokinetic* sampling such that the streamlines are undisturbed and the instrument neither underreads nor overreads (figure 4.18(*a*)). If, however, the gas velocity is too low (figure 4.18(*b*)), the flow will diverge around the sampling tube, hence giving an underestimate. (Figure 4.18(*c*) shows the overestimation effect resulting from an excess velocity.) Dusts and gases can also be withdrawn for further analysis by this method. Such monitoring is important for maintaining compliance with emission standards and/or 'best practicable means' (see Chapter 8). Statistics for emissions have been presented briefly in Chapter 1. Any pollutant thus added to the atmosphere will result in an increase in both atmospheric and ground level concentrations.

OBSERVED POLLUTANT LEVELS

Presentation of statistics for ambient concentrations of atmospheric pollutants can be made in tabular or graphical format on a variety of time and space scales (e.g. continuous, mean daily, annual average, point, average city, average country, global). The difference effected by varying the time averaging period has been illustrated in figure 4.8. In this section some recent data for the UK and the USA are given on national and regional scales.

Many of the ambient pollution statistics are expressed in terms of data derived from integrated daily samples (obtained by means of the SO_2 bubbler for example). Indeed the World Health Organisation (WHO) targets are defined in these terms: $40 \, \mu g m^{-3}$ (smoke) and $60 \, \mu g m^{-3}$ (SO_2). Figure 4.19 shows such daily data (averaged over each monitoring year which, in the UK, runs from April to March) for smoke and SO_2 respectively and table 4.1 shows data for several EEC countries for SO_2 for 1977. Over the UK as a whole, the WHO target has been met for smoke but not yet for sulphur dioxide. Spatial variations are however great—the urban areas being above this average in general. Figure 4.20 shows the analogous data (SO_2 only) for Manchester and Salford for the period April 1974 to March 1980. Data are shown not only for the annual mean values but also for winter (upper curves) and summer (lower curves). A decreasing trend is observable in both sets of data, but the areas do not as yet attain the WHO objective. They are delimited as areas at risk of violating the new EEC standards (see Chapter 8) in a recent (June 1981) assessment by the Warren Spring Laboratory. Although smoke levels in London for 1980–81 were (perhaps anomalously) low and below the EEC limits, sampling stations have historically been sited away from street level. Pilot studies suggest that street-level smoke concentrations may be locally up to five times higher than those reported in the National Air Pollution Survey. These data sets, together with statistics for other areas, suggest that there has been a marked decrease in SO_2 and smoke from the mid 1950s until the late 1960s followed by a less rapid decrease. The decrease has been attributed to Clean Air legislation, an increasing trend to central heating and away from domestic open fires and anomalous meteorological conditions. No doubt all have contributed to varying degrees, although this success in alleviating atmospheric pollution has resulted in some complacency and the pressure for instigating Smoke Control areas is becoming less.

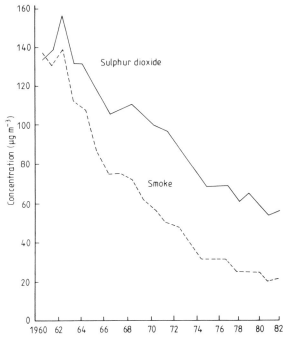

Figure 4.19 Trends in average urban concentrations of smoke and SO_2 in the UK. Data from *Warren Spring Laboratory Reports*.

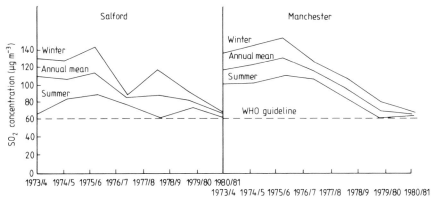

Figure 4.20 Trends in SO_2 concentrations for two urban areas in the UK (Salford and Manchester). Annual, winter and summer mean values are shown (years run from April to March).

Table 4.1 Mean sulphur dioxide concentrations in 1977 for EEC countries, using acid titration method for evaluation. A = towns with populations greater than 500,000; B = towns with populations less than 500,000.

Country	Class	Number of sites	Mean (μg m^{-3})	Highest daily value (μg m^{-3})
Belgium	A	11	84.9	378.5
	B	24	77.6	315.7
Denmark	A	5	41.8	285.6
France	A	35	63.5	322.4
	B	33	56.2	496.4
Ireland	A	4	34.0	252.0
	B	2	24.0	98.0
Luxembourg	B	4	35.5	190.8
United Kingdom	A	41	80.6	389.7
	B	30	54.4	225.8
All countries	A	96	70.9	352.7
	B	93	59.6	340.8

For the USA national figures have little significance. Temporal decreases are reflected in the increasingly stringent permitted ambient concentrations (see Chapter 8). These are given in terms of values not to be exceeded more than once per year. Observed maximum 24 hour averages for urban areas indicate the variability that occurs and the difficulties of presenting data in a meaningful way.

The mean daily approach may conceal short periods of high concentration which may have deleterious effects on health (see Chapter 5). Indeed mean daily values of 100 μg m^{-3} of SO_2 are not inconsistent with short period peaks of over 1000 μg m^{-3}. Diel variability of pollutants may be attributed to many factors including traffic (see also Chapter 7), meteorology and level of industry. In Cairo (figure 4.21) it has been shown (Abdel Salam *et al* 1981) that smoke concentrations were higher in the winter months as a result of a higher degree of atmospheric stability and that daily minima occurred in the early afternoon when highly turbulent (convective) motions dispersed the smoke over a greater volume. However it is the extended periods of high concentrations that are most damaging. These are usually the result of persistent anticyclonic (northern hemisphere) conditions (see Chapter 2) coupled with high emission rates (e.g. in winter). Such a pollution period is known as an *episode*. Such occurrences have been well documented and were the spur for Clean Air legislation in the UK after the disastrous smog in London in December 1952. Some estimates of 24 hour SO_2 concentrations for typical episodic conditions are given in table 4.2.

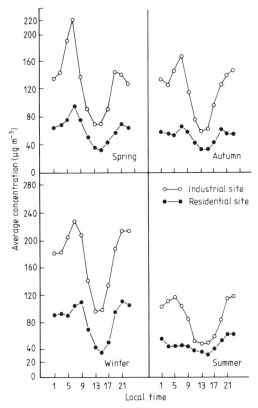

Figure 4.21 Seasonal diel variation of smoke concentration in the Greater Cairo area (June 1977–May 1978). Redrawn from Abdel Salam *et al* (1981) with permission.

Table 4.2 Some episodic events and their associated SO_2 levels.

Episode	Date	Estimated concentration $SO_2(\mu g\,m^{-3})$
Seraing (Meuse Valley)	Dec 1930	25000 (estimated value)
Donora, Pennsylvania	Oct 1948	1600 (estimated value)
London	Dec 1952	3830
London	Dec 1962	3300
New York	Mar 1964	1730
New Jersey	Sept 1971	~ 0.1 ppm oxidants

SUMMARY

Pollutant monitoring, especially the design of a monitoring network, is of much current interest and concern. Sampling programmes are undertaken in two (horizontal) and sometimes three (also vertically in the atmosphere) spatial dimensions over periods of time; the length of the programme being determined by the programme objectives and availability of resources. It is essential that the results obtained are statistically sound and worth the monetary investment they require.

The sampling train has four identifiable portions: inlet, measuring device, meter and air mover. Both laboratory and on-site measurements have been discussed. Whether to sample for ambient concentrations or sample at source (e.g. in the ducting or chimney) will depend partly on the party undertaking the monitoring as well as concern to meet legislative requirements. Values for smoke and SO_2 for Europe and the USA have been given as well as some indication of the guidelines different countries are currently aiming to meet.

5
Effects of Pollutants

Damage caused by air pollution can be assessed in terms of effects on health and deterioration of inert materials (especially buildings), plants, animals and degradation of the atmosphere itself. Costings are difficult. How can one assess the cost of the personal injury resulting from a lung disease? The cost of medical treatment is only a part of the adverse effect especially if the disease can only be controlled and never totally cured. Damage to stonework can be evaluated by the cleaning or replacement cost, but for art treasures replacement is not even feasible.

Figure 5.1 Writing pads. The right-hand one is new. The left-hand one has been on a desk exposed to sunlight (in the South African winter) for a period of seven days. Marked fading (colour loss) is evident.

A corrosive atmosphere (i.e. one containing damaging pollutants) may be acidic or alkaline, containing oxidants, salts and organic or biological material. Some materials deteriorate even in a 'crystal-clear' atmosphere due to the effects of sunlight e.g. sunlight weakens fabric and fades and discolours paper (figure 5.1). High humidity, alternate freeze–thaw mechanisms, natural dusts and sands can all be responsible for deterioration. Degradation processes are extremely effective over geological time periods, soil being largely a result of natural weathering. Damage may result from a short exposure to a high concentration and/or a long exposure to a lower concentration. It is nevertheless possible to assess, at least qualitatively, the deleterious effects of anthropogenic pollution on both inert and living material.

INERT MATERIALS

Table 5.1. summarises some of the more important deteriorations observed in inert materials.

Table 5.1 Deterioration of some materials in a corrosive atmosphere

Material	Deterioration observed	Measurement
Metal	Spoilage of surface, loss	Reflectance, weight change
Building stone	Discolouration, leaching	Not quantitative
Fabrics, dyes	Discolouration, fading, weakening	Reflectance, loss of tensile strength
Leather	Weakening, embrittlement	Loss of tensile strength
Paper	Embrittlement	Decreased folding resistance
Paint	Discolouration	Reflectance
Rubber	Cracking	Loss of elasticity, cracks

Metals

Metals may suffer corrosion directly either by acid or alkali. Pollutants removed from the atmosphere by rain or washout may overlay the metal and form an electrolytic cell in which the pollutant acts as the electrolyte. For example, a zinc surface will release electrons into solution which combine with hydrogen ions available in an acidic electrolyte to liberate gaseous hydrogen.

$$\text{Anode:} \qquad \text{Zn} \longrightarrow \text{Zn}^{2+} + 2e$$

$$\text{Cathode:} \qquad 2\text{H}^+ + 2e \longrightarrow \text{H}_2$$

$$\text{In summary: } \text{H}_2\text{SO}_4 + \text{Zn} \longrightarrow \text{ZnSO}_4 + \text{H}_2.$$

Thus the zinc anode is corroded. Coating with steel may help in the case of zinc and aluminium since the steel forms the cathode and hence does not suffer corrosion.

Alternatively protection is possible by painting or by the addition of other elements. (For instance, carbon steel which contains 0.02 % Cu is less resistant than steel containing 0.2 % Cu plus traces of nickel and chromium.) Some metals are easily oxidised (e.g. Al to Al_2O_3). A metal surface will thus readily coat itself with a protective layer of the oxide thus preventing the further contact of the air pollutant with the underlying metal.

In many of these processes, the moisture content of the atmosphere is critical. Corrosion is relatively slow when the relative humidity drops below 60 %. As the humidity approaches 80 % there is a marked increase in corrosion which continues to rise with increasing relative humidity—although during periods of rain, corrosion is minimal. Temperature also has a direct effect, corrosion being more rapid at higher temperatures. Thus cool, dry climates provide the best environment for metals. In countries such as South Africa, where humidity is low and temperatures high, corrosion to vehicles is noticeably less.

Building stone

Building stone, such as limestone, containing high percentages of carbonates (e.g. of calcium and magnesium) is very susceptible to the attack of acidic pollutants. The carbonate is changed to sulphate—which is relatively soluble and readily leached away. In addition the greater volume of the sulphate helps to fragment the limestone by inducing a mechanical stress. The effects of SO_2 and CO_2 in solution (i.e. carbonic acid) are:

$$CaCO_3 + SO_2 + \tfrac{1}{2}O_2 + H_2O \longrightarrow CaSO_4 + H_2O + CO_2$$

$$CaCO_3 + CO_2 + H_2O \longrightarrow Ca(HCO_3)_2$$

$$Ca(HCO_3)_2 + SO_2 + \tfrac{1}{2}O_2 \longrightarrow CaSO_4 + 2CO_2 + H_2O.$$

Building stones such as gneiss, granite and sandstone which do not contain substantial amounts of carbonates, are more resistant to attack by acidic gases. Bricks are resistant to sulphur dioxide and sulphuric acid, but the mortar used as a bond is not, hence the frequent need for repointing of mortar.

Fabrics and dyes

Deterioration of fabrics is perhaps one of the more readily discernible effects of air pollution. Fading and soiling of clothes is not only of an aesthetic concern but also leads to the necessity for more frequent cleaning (and of course the greater economic stress of large cleaning bills). This deterioration is effected by particulates, acid gases (e.g. SO_2) and oxidants (e.g. O_3, NO_2, PAN) and adds to the biological spoiling and the effects of sunlight and humidity. Many dye molecules are activated by sunlight which may result in fading or gradual colour changes; for instance blue disperse dyes and violet dyes tend to redden under the influence of oxides of nitrogen.

Paper, leather and paint

For paper and leather the effects of pollutants are seen in a more rapid tendency towards a brittle state, although such effects are less indoors. Both paper and leather adsorb SO_2. With high humidity conditions, concentrations of 10 ppm of SO_2 cause leather to rot within about six weeks, although some protection is afforded by treating with a solution of potassium lactate.

Although designed to protect as well as decorate, paint itself can be subject to air pollutants. High concentrations (of the order of 1–2 ppm) of SO_2 can increase drying times and exposure of lead-based paint to H_2S leads to a rapid darkening as black lead sulphide is formed by the reaction

$$Pb^{2+} + H_2S \longrightarrow PbS + 2H^+.$$

The rate of this reaction depends upon the concentration of H_2S. In time lead sulphide oxidises so that the paint regains its original colour.

The impact of particulates on a painted surface may cause pitting and weakening of the protective surface. Frequent washing may exaggerate this, whilst increasing the aesthetic appeal of a newly cleaned surface. Frequency of repainting (of external woodwork) is found to be well correlated with the prevailing concentration of deposited particulates.

Rubber and other materials

It has long been known that rubber is cracked by ozone, which attacks the double carbon bond. Since the presence of ozone in the lower troposphere is indicative of photochemical reactions, this problem is largely confined to areas of high irradiation (such as the west coast of America). Natural rubber and butadiene–styrene rubber are susceptible to ozone because of the double bonds (in the small percentage of dienes present) and are thus relatively susceptible to degradation. Silicon rubber is, on the other hand, saturated—the absence of double bonds ensures its resistance to the attacks of ozone.

Oil paintings have suffered, over years of display, from deposition, irradiation and acid gas attack. Television tubes have been found to have a shortened life when used in a polluted environment and dyed hair has been discoloured by the components of the photochemical smogs prevalent in cities such as Los Angeles.

VEGETATION

Optimum plant growth requires adequate light, heat, moisture, nutrients and appropriate soil conditions. An imbalance in any of these results in a stress to the plant (see table 5.2) which may result in restricted growth or foliage markings. Pollution provides an extra undesirable stress. If this stress is too high, then the plant will die, despite the relatively complex biological defence mechanisms (e.g. rebuilding of damaged tissue).

Autotrophic plants take in air and energy (sunlight) and through the process of photosynthesis use them to produce carbohydrates, releasing oxygen as a byproduct. The leaves, containing the chloroplasts used in photosynthesis, are an important part of the plant and a major growth area. It is through the leaves that a plant may absorb a gaseous pollutant, because one of their major functions is to absorb atmospheric gases. Examination of leaf structure reveals three regions: the epidermis (outer layer), the mesophyll and the veins which transport water and nutrients around the plant. Gases and small particulates enter the leaf through the stomata (leaf pores), to the air spaces of the mesophyll.

Table 5.2 Pollutant stress factors affecting plant physiology (source Stern *et al* 1973).

Factor	Effect
Light	Pollutant injury generally decreases with increased light intensity (PAN has the opposite effect).
Exposure	Dosage for damage/injury species-dependent. Fumigation frequency additional harmful factor.
Relative humidity	High sensitivity to pollutants at 100% RH. Reduced sensitivity below 50% RH.
Nutrients	High nutrient levels (maximum growth) increases susceptibility to pollutants.
Temperature	Increased injury at higher temperatures (i.e. optimum growing conditions).
Moisture availability	Probably most important factor for cell physiology. Plants near wilting conditions resistant to pollutants.
Seasonal-diel variation	More resistant at night (stomata closed). Injury more severe in spring and early summer than in autumn and winter.

The precise mechanisms of damage by pollutants within the leaf are not well understood. Indeed low quantities of SO_2 are vital to plants for good growth. This is converted to sulphate ions and it appears that only when the $SH:SO_4$ ratio exceeds some threshold value or when SO_4^{2-} ions accumulate, does damage occur. Symptoms of such damage include

(*a*) necrosis and bleaching of leaf margins;

(*b*) glazing and silvering of surfaces, especially the undersides;

(*c*) chlorosis (loss of chlorophyll);

(*d*) flecking or stippling of upper surfaces.

However it is important to note that all the above effects can result from natural factors, e.g. abnormal temperatures, insufficient water.

Reducing pollutants such as SO_2 act on the mesophyll cells creating necrotic areas between the veins, breaking down chloroplast membranes and bleaching protoplasts.

Oxidants such as O_3 create local cellular collapse and pigmentation of the cell walls, with possible lesions on the upper surfaces. It has been established (1982 figures) that in the USA ozone could be responsible for over 3×10^9 each year in terms of damage to crops. Experiments have demonstrated that the exposure of a peanut crop to 0.120 ppm (compared with a natural background of 0.025 ppm) decreases the yield by 50%. (Present ambient levels are in the region of 0.05–0.07 ppm over all the USA except California where levels are noticeably higher.)

The reaction of the plant to a given pollutant also depends on the plant species. It is vital to use a control group for comparison in any experiment used to evaluate the effects of air pollutants.

A phenomenon which has been recently observed in plants is *synergism*. This term describes a situation in which the damaging effect of two or more pollutants is greater than the sum of the damage of the individual pollutants. Using various grasses it has been shown that grass exposed to a combination of sulphur dioxide and nitrogen oxides is stunted far more than following sequential exposure to individual gases of the same concentration. Similarly the effects of SO_2 appear to be greatly exacerbated in the presence of fluorides.

Although *damage* to plants has been discussed, this term should be restricted to that pollutant effect which results in either aesthetic or economic loss. Economic losses can occur not only when the crop is destroyed but also when growth is inhibited. For example, exposure to a pollutant may result in stomatal closure and hence in slowed growth rates. However this action helps the plant to avoid injury whilst the maturation period will increase. In contrast to damage, the term *injury* is applied when a part of the plant dies but the overall growth of the plant is not affected (e.g. the effect of one drop of sulphuric acid applied to the plant would be local only).

ANIMALS

Experimental

The tolerance of different species of animals to pollutants is markedly different. Both long and short term exposure to pollutants may have immediate and/or mutagenic effects. For example many larger birds have over the last few decades neared extinction as a result of widespread use of DDT (dichlorodiphenyltrichloroethane) as a pesticide (insecticide). DDT accumulates in body fats; in birds this results in thinner eggshells, which are thus more fragile and susceptible to accidental damage. This is a phenomenon which is observed worldwide, e.g. among the Adélie and Emperor penguins of the Antarctic. In America, human milk often contains a greater amount of DDT than that permitted in powdered milk available on the retail market.

The effects of pollutants are seen also in the *melanism* of, for instance, the peppered moth. In northern England, soot covered trees and walls rendered useless the brown camouflage of the moth (see figure 1.2). Instead the black strain thrived since it was better adapted to polluted surroundings (being protected by its black colouring from predators)—happily the original strain is reestablishing itself in those areas colonised until fairly recently by the black variety. Adaptation to changing pollution levels is also seen in reinvasion of many urban areas by lichen species as a direct result of the ameliorating atmospheric conditions.

Experimental animals may be used to evaluate the effects of pollutants, and results so obtained can be 'scaled up' and applied to man. The largest single obstacle to the efficacy of the experimental procedure is that of extrapolation, in other words assuming that a body weight ratio scaling of toxic animal doses to man is valid. The biologies of say a mouse and man *may* react identically (taking the size-scaling into account) but there is no direct proof other than the test of time. Safety is assumed if the test animal survives doses many times larger (*pro rata*) than the human population would be likely to be exposed to. In respect of exposure to specific air pollutants, animal experiments have been (and/or are being) conducted using in particular ozone, nitrogen oxides, sulphur dioxide, carbon monoxide and particulates. The effects of *ozone* on animals are similar to those of ionising radiation, including possible damage to chromosomes.

The effects of ozone are often roughly proportional to dosage (concentration × exposure time) i.e. there is a no-threshold dose response. The olfactory nerve endings are damaged leading to premature aging and chromosomal injury. A threshold value of 0.1 ppm is often quoted as applicable. On the other hand, it appears that a critical concentration of NO_x is needed before any ill-effects are noted. The major effect is likely to be damage to the lungs, possibly resulting in death. Although the industrial threshold of NO_2 is set at 5 ppm, rabbits exposed to concentrations as low as 1 ppm over a period of one hour have suffered protein changes.

The toxicity of *sulphur dioxide* is a matter of concern to the population of the industrialised world. Experiments show a wide variability in the degree of susceptibility of different species and different individuals. At concentrations of 1 ppm for a year there is a marked increase in animal mortality (epidemiologically observed effects in humans are discussed later in this chapter). This acid gas acts as an irritant to the respiratory tract, especially to the mucous lining.

Carbon monoxide, a product of partial combustion, is a known toxin. It diffuses from the lungs to the blood stream where the haemoglobin has an affinity for CO 210 times stronger than that for O_2. The oxygen carrying capacity of the blood is thus reduced, and if no oxygen is being delivered to the cell site mortality rapidly results. The symptoms are related to the ratio of $HbCO : HbO_2$ (Hb denotes the haemoglobin radical) which should be kept below about 5 %.

Acid particles such as sulphuric acid, ammonium sulphate, hydrochloric acid and hydrofluoric acid can exert a direct chemical action; as can bases such as sodium hydroxide, calcium hydroxide and ammonium hydroxide. Inert particulates, such as fly ash, carbon, iron and asbestos, can be responsible for slowing the ciliary beat (see the section on the respiratory tract) and reducing mucus flow in the bronchial tract. *Synergism* is again an important research topic. It is vital to recognise combinations of pollutants that may have a synergistic effect—animal investigations may once again assist here.

Commercial

Damage (both physical and economic) to commercial animal stocks is seldom a result of exposure to contaminated air. More frequently it occurs as a result of two steps: firstly an accumulation of airborne pollutants on vegetation later to be used as foodstuff and secondly ingestion of the contaminated food. Herbivorous animals may suffer accumulations of poisonous elements within their bodies; whilst carnivores receive their poison pre-concentrated as a result of their position higher in the food chain.

Much of the damage to commercial animals is a result of metal poisoning. Metals such as lead, molybdenum, mercury and arsenic (which possesses both metallic and non-metallic properties) have been responsible for the loss of large numbers of sheep and cattle. Arsenic and lead originate from smelters. Arsenic, present in food as arsenic trioxide at a concentration of 10 mg As per kg of body weight, leads to severe colic and cirrhosis of the liver and spleen in cattle. Molybdenum arises from steel plants and some roasting processes: 230 mg per kg in feedstock is considered dangerous. Mercury is a waste product of chlorine-caustic plants—fish containing more than 0.5 mg kg^{-1} of mercury is legally unsaleable in many countries. $25-50$ mg kg^{-1} of lead in cattle causes frothing at the mouth, muscular spasms and paralysis of the larynx muscles. Gaseous emissions can also be a hazard. Some phosphate fertilizer plants emit fluorides and extensive exposure may result in fluorosis: teeth and bones are eroded resulting in stiffness and lameness in the animal.

HUMANS

It is important to ask whether any specific pollutant will have harmful effects on man. A positive indication of harmful effects is only acquired after the onset of disease in or on death of an individual because experimental work on humans is not ethical. *Epidemiological studies* require long term observation of sample populations with similar exposure over a number of years and comparison with a *control group,* if this is possible. Most conclusions (e.g. regarding the danger of smoking) are based on such studies. The use of animals for such research has often replaced controlled tests on humans, because it is assumed that the loss of animal life in the laboratory is greatly preferable to endangering a large proportion of our own species on a large scale, for example by release of a dangerous chemical into the water supply or atmosphere.

Detrimental effects of pollutants on humans have been observed, if not always predicted. Many effects are not immediately obvious as a result of the efficiency of the body's response to stimuli as it attempts to retain internal stability (or homeostasis). Small perturbations (or *stresses*) are easily overcome. However the reaction depends on other stresses that may exist and indeed whether there is a synergism. Social stresses (such as late nights, anxieties about work, marital problems) may compound the effect of the pollutant; as can the physical state of health of the subject. Those especially vulnerable to acute effects are the old, the young, the pregnant and the ill. The spread of response is wide, making determination of ill-effect/no effect and safe/dangerous decisions very difficult. Even within a strong and healthy population, individuals show a markedly different capacity to recover, and many of the cause–effect links are only suspected, not proven. Indeed when the effect of the pollutant is to cause poor health (or worse), proof is seldom sought and inference and extrapolation techniques are used extensively.

Humans can respond to a pollutant either as a result of its visible or olfactory effects or its pathogenic or potentially pathogenic effects. Pathogenicity can result from the immediate toxicity of the pollutant or from less well understood long-term effects. The general population may often not be able to specify the 'problem pollutant' because the most immediately obvious pollutant (for example a particulary odourous chemical) may not be the one having the most harmful potential. Response may be measured for an individual or (statistically) for a population in terms of the physiological sensations of sight, smell, touch and taste or the medical effects.

An idealised response curve for an individual is shown in figure 5.2: no response below the threshold and 100 % response at greater concentrations. Many atmospheric pollutants can be considered as having this well defined threshold—but this threshold value is not identical for all members of a population. A threshold distribution for the population is more likely to be normal or lognormal (figure 5.3). In the former, the largest group of people have an average response, with smaller numbers towards both the insensitive and supersensitive ranges. The lognormal distribution has a median (middle value) and mean (arithmetic average) to the right (in the example shown in figure 5.3) of the maximum value (or mode). A better representation is given by a cumulative distribution (figure 5.4). In this S-shaped (logistic) curve, the percentage of the population responding to a given concentration (or below) is plotted. Thus for a population it is more difficult to identify a 'safe' concentration. A threshold may be associated with the rapidly rising portion of the curve and this permits a legally enforceable limit to be agreed upon; a value which will nevertheless leave a small percentage of the population susceptible to harmful effects.

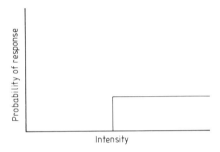

Figure 5.2 Step function response curve for an idealised situation.

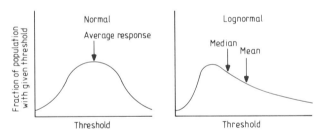

Figure 5.3 Normal and lognormal distributions.

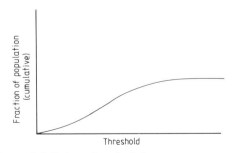

Figure 5.4 S-shaped or logistic response curve.

Notwithstanding, the American Conference of Governmental Industrial Hygienists regularly issues a set of guidelines for permissible exposures over a forty hour working week. These give a *threshold limit value* (TLV) for a wide range of substances (see table 5.3) as well as figures for shorter term (up to fifteen minutes) exposure and ceiling values (which must never be exceeded even instantaneously) for chemicals having a toxic effect after a very short exposure (e.g. gaseous irritants). Although specifically designed for the working environment, for ambient atmospheric concentrations many pollution scientists use the value of TLV/40 as a rough guideline for permissible exposures for the general public. Nonhebel (1981) suggests for combustion generated pollutants other than SO_2, that a three minute average of less than TLV/30 should be used as a threshold value.

Table 5.3 Examples of threshold limit values (TLVs) taken from the Health and Safety Executive Guideline Note EH 40/84. Reproduced by permission of the Controller of Her Majesty's Stationery Office.

Substance	TWA†		STEL‡	
	(ppm)	(mg m^{-3})	(ppm)	(mg m^{-3})
Acrolein	0.1	0.25	0.3	0.8
Ammonia	25	18	35	27
Carbon dioxide	5000	9000	15000	27000
Carbon monoxide	50	55	400	440
Fluoride (as F)	—	2.5	—	—
Nitrogen dioxide	5	9	5	9
Nitric oxide	25	30	35	45
Sulphur dioxide	2	5	5	13

† TWA = Time weighted average.
‡ STEL = Short term exposure limit.

The potential for a biological response may be characterised by several or all of the following:
(*a*) ambient concentration;
(*b*) period of exposure;
(*c*) lifetime dosage (dose = concentration × exposure time);
(*d*) dosage during limited time;
(*e*) dosage received by a certain percentage of the population.
Figure 5.5 is an attempt to relate graphically the possible health damage caused by SO_2 in terms of concentration and exposure time. It is evident that at high concentrations, of say 1 ppm, an exposure of less than one hour can cause some damage; whereas at 0.01 ppm, 6–9 months is required for any effect. If these figures were expressed as dosages, then the range would be from about 1 (high concentrations) to 80 (low concentrations) (units

ppm × h). It has been shown that there is a good correlation between dose and exposure period. Some observed doses and their average duration for the Milan area of Italy are given in table 5.4, although the importance of the choice of averaging period for monitoring must be stressed in order that high doses of short duration will be recorded. Interpretation of figures thus requires care and such figures should not be used as a single criterion for inferring the potential for human response to airborne pollutants.

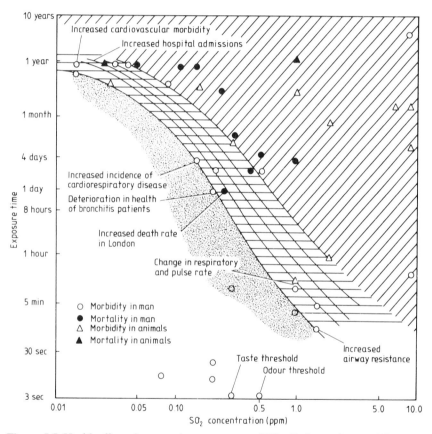

Figure 5.5 Health effects due to various exposures of SO₂. Redrawn from Williamson (1973).

The 'points of entry' for pollutants to the body are in general few. On average we spend our lives with about 90 % of our body surface clothed; leaving in contact with the atmosphere, face, hair, hands, nails and teeth. These exposed parts may be affected directly e.g. some gases and partic-ulates cause lachrymation. Tears thus shed are effective in removing the original stimulus from the eye and thus the response is temporary and the problem rapidly alleviated. Exposed dead tissue (nails, hair) is seldom affected directly, although ingestion of arsenic, selenium and fluorides may first become evident in such areas.

Table 5.4 Dosage statistics for SO_2 data for the Milan area (after Drufuca *et al* 1980).

Dose (D) (ppm × h)	Probability $p(D)$	Number of events	Average duration (h)
$D < 0.1$	0.62	68	0.8
$0.1 \leq D < 0.2$	0.06	6	2.1
$0.2 \leq D < 0.4$	0.12	13	3.0
$0.4 \leq D < 0.8$	0.07	8	5.2
$0.8 \leq D < 1.6$	0.10	11	6.4
$1.6 \leq D < 3.2$	0.03	3	9.9

In some cases where pollutants are suspected of causing damage to health the disease is a direct result of exposure to polluted conditions; in others the pollutant is one of several stresses initiating the disease. Sometimes the pollutant will in no way be involved in the initiation, but strongly implicated in exacerbating the disease once established (see e.g. discussion by Waller 1983). It is however clear that certain diseases (notably respiratory diseases, see below) are linked in some way to pollutant levels. It is thus important to understand the basic working of the (human) respiratory system.

The respiratory tract
Figure 5.6 shows the air route in the respiratory tract. Air is inhaled through the nose, wherein lies the first line of defence against airborne particulates. Hairs in the nose filter out dust greater than about 10 μm in size (dependent on dust loading in the atmosphere). The trachea (and all other parts of the bronchial tract *except* the alveoli) are lined with slowly oscillating hairs called *cilia*. Particles impinging on the wall sides are thus trapped both by the cilia and by mucus secretions and slowly advected to the top of the oesophagus, where they are then swallowed and eventually excreted. The trachea divides at the top of the chest into two bronchi which themselves subdivide (about a further 20 times) into bronchioles of decreasing size until reaching the air-sacs or alveoli. There are about 3×10^8 alveoli each about 0.2 mm in diameter. The consequent surface area, across which oxygen diffuses into the blood and carbon dioxide out of it, is about 50 m^2 (compared with an average outer skin area of only 2 m^2).
Removal of particulates within the bronchial tract depends upon particle

size. Those less than about 10 μm may pass through the nose and be collected in the bronchial airways, although particles as small as 1 or 2 μm, which exhibit Brownian motion, may not impinge on the wall sides and will thus enter the alveoli. Unfortunately there are no cilia in the alveoli. The removal mechanism here is either (for soluble particles) through the alveolar walls into the lymph system or blood stream or by means of macrophages (mobile single cells) which engulf the dust and eventually die and, by random advective motions, are deposited at the lower end of the bronchiole from where the cilia are able to remove the particle.

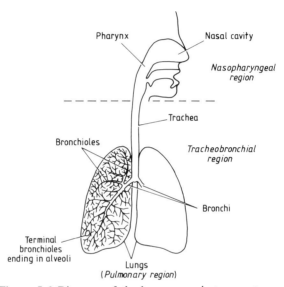

Figure 5.6 Diagram of the human respiratory system.

Figure 5.7 shows that larger particles are almost totally removed in the nasopharyngeal region, but all particles smaller than about 10 μm pass through unimpeded. These smaller particles are removed with an efficiency increasing with decreasing size in the tracheo-bronchial and pulmonary regions. The net result is that for particulates greater than about 0.1 μm and less than about 10 μm, the efficiency of removal is very low (less than 50 %) such that this size range is likely to be most damaging to the lungs.

Respiratory illnesses are correlated with the amount and location of dust deposition within the lungs, especially in the alveoli. Mucus secretions from the bronchiole walls exacerbate phlegm formation, irritating the tract and resulting in excess coughing. For a typical (clean air) dust content of say $0.01 \, \text{mg m}^{-3}$, one can calculate roughly the daily dust intake independently of size. Daily air inhalation is on average about 13 kg or approximately $7.6 \, \text{m}^3$. Hence we may inhale, on average, almost 0.1 mg of dust each day.

If only a very small percentage of this dust penetrates into the alveoli, respiratory illness may result, particularly if the dust particles are irregularly shaped (e.g. asbestos dusts) and especially if the exposure is of extended duration. Many respiratory diseases are associated with specific occupations—their name often reflects this.

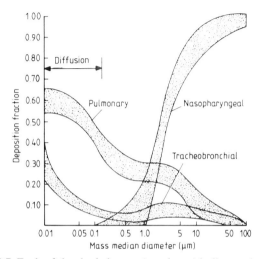

Figure 5.7 Each of the shaded areas (envelopes) indicates the variability of deposition for a given mass median (aerodynamic) diameter in each region of the respiratory system. Redrawn from Air Quality Criterion for Particulate Matter *National Air Pollution Control Administration Publication AP-49* (1969) (Washington, DC: US Govt. Printing Office).

Bronchitis was at one time called the 'English disease' as a result of the high dust loading in a damp, foggy atmosphere. It is characterised by inflammation of all or part of the bronchial tract resulting in excess mucus production and initiating a cough reflex. The airway becomes clogged by the mucus, decreasing its internal cross sectional area and hence the airway resistance is increased and breathing becomes more difficult.

Pulmonary emphysema results from a thinning of the alveolar walls and is a progressive disease with no known cure which is fast becoming a major cause of death in the developed world. US figures show 17 deaths in 1950 from emphysema rising to 20 252 in 1966 (although some of the difference may of course be due to increased diagnostic facility and increased population).

Lung cancer develops in the epithelial lining of the bronchi resulting in malignant tumours which block the airway. A major initiating factor in the development of lung cancer is probably cigarette smoking—it has been established that some components of cigarette smoke (e.g. polycyclic aromatic hydrocarbons) are carcinogenic. Concentrations may be more than 10 times higher than in unpolluted air; although the carcinogenic properties of 'sidesmoke' (i.e. that produced by a cigarette but not inhaled by the smoker) mean that non-smokers suffering discomfort in a room full of smoke may have a medical basis for their disquiet. In China scientists have shown that a strong correlation exists between incidence of lung cancer and cooking over open coal fires. In certain rural areas, where women rarely smoke but perform most of the cooking, cancer deaths are noticeably higher in women than in men.

The incidence of *mesothelioma* (cancer of the lining of the lung) is closely linked with exposure to asbestos and is prevalent among workers in the asbestos industry and among other workers having close contact with asbestos (Gardner *et al* 1982).

A group of lung diseases resulting from dust inhalation can be referred to by the general name *pneumoconiosis*—these can include silicosis (for example quartz and sand as found in foundries, potteries and in sand-blasting), anthracosis or coal miner's 'black lung' (from coal dust), asbestosis etc. Such diseases result in a progressive inflammation of the lungs causing irreparable damage.

People often do not react to air pollutants merely out of concern about possible health hazards. Often the primary reaction is related to the smell or appearance of the pollutant. There is sometimes but not always, a correlation between 'unpleasantness' and the effects on health. The senses most often affected are, probably in order, smell (olfactory sense), sight and taste.

Odours can be classified by their quality and their intensity. It must be stated that assessing the quality of an odour is a subjective matter. What is pleasant to one person may be unpleasant or oppressive to another, and pleasantness or unpleasantness can often be as a result of a mental association (e.g. the hop smell from breweries or freshly ground coffee). The true mechanism of smell is unknown, although it is thought that the olfactory organ is an area of 2.5 cm^2 situated in the roof of the upper nasal cavity. Although the odour quality often seems to depend on the shape of the molecule, there is as yet no way of predicting the odour of a molecular species by assessing its shape. In a mixture of odours one will often dominate. This is used to good effect in deodorants and air fresheners where an unpleasant smell is *masked* by another odour.

We can become aware of an odour occurring in two ways. At a certain concentration an odour is detected, although identification is not possible. Often an increase in the concentration by a factor of ten is needed before the odour is identified, although whether or not an odour is identifiable depends on prior experience with that chemical. As discussed earlier in this chapter, any individual will have his or her own personal *detection threshold* and *recognition threshold*. For a population this leads to two response curves. Published threshold concentrations are thus only indicative of a recognition threshold. These are usually derived by a panel of four trained observers (the only reliable instrument is still the nose) under strict conditions. It is known, however, that extended exposure to an odour can lead to an increase in the recognition threshold. Some typical odour recognition thresholds are given in table 5.5.

Table 5.5 Odour recognition threshold values (ppm) (data from Leonardos *et al* 1969).

Chemical	Threshold
Acetic acid	1.0
Acetone	100.0
Acrolein	0.21
Ammonia	46.8
Chlorine	0.314
Formaldehyde	1.0
Hydrogen sulphide	0.00047
Sulphur dioxide	0.47
Trimethyl amine	0.00021

Many sulphur compounds have low thresholds e.g. methyl mercaptan (CH_3SH) and ethyl mercaptan (C_2H_6SH). Hydrogen sulphide (H_2S) is usually recognised at concentrations between 1 ppb (most sensitive) to 10 ppb (least sensitive). Not all low threshold compounds are malodourous, e.g. artificial musk can be recognised at a concentration of 0.007 ppb. Odours can often be controlled by dilution, treatment by incineration, absorption, adsorption or oxidation (see Chapter 8).

Visibility may be reduced by atmospheric pollutants thus affecting the sense of sight. Before the Clean Air Acts (UK and USA) the common appreciation of the existence of pollution was by sight. Visibility is reduced by the presence of a pollutant aerosol, which absorbs and scatters light (and may take part in complex photochemical reactions).

Scattering is size dependent, although most aerosols are less than 10 μm in size (either primary or secondary pollutants). The type of scattering depends on the ratio of the particle size to the wavelength of light (visible light wavelengths are in the range of 0.4–0.7 μm). Scattering is accomplished by a combination of reflection, diffraction and refraction—the incoming light is thus deflected away from its original path and away from the eye of the observer. The aerosols which have the greatest scattering effect are those of a size approximately equal to the wavelength of light. Shorter wavelength light (e.g. blue) is scattered more. (It is for this reason that sunsets are red, since the blue light has been removed from the light beam. This occurs because the path length through the atmosphere, where all the scattering occurs, is greater than at any other time of the day.)

Calculation of scattering coefficients depends upon the size of the scatterer, its refractive index and absorptivity. Rayleigh scattering describes that resulting from the presence of particles very much smaller than the wavelength of light (say much less than 0.5 μm); Mie scattering is used for particles of the same size as the wavelength of the radiation.

Measurement of the visibility is accomplished in terms of the *visual range*—the largest distance at which objects can be discerned. Koschmieder (1924) derived the following expression for the range. Viewing a black target on a white background, the contrast perceived at distance x, $C(x)$, is the apparent luminance given by

$$C(x) = [I_2(x) - I_1(x)]/I_2(x) \qquad (5.1)$$

where I_1 and I_2 are the intensities from the target and background respectively. I_1 is determined by the absorption of the light in the atmosphere plus any extraneous scattered light. Assuming that the atmosphere is homogeneously polluted, then we consider a cylinder of light as it passes through the atmosphere. The beam is depleted by aerosol absorption by $k\Delta x I_1(x)$, and by scattering by $b\Delta x I_1(x)$ and is augmented by scattering into the beam by $b'\Delta x I_0$ where k, b, b' are empirical coefficients. Thus

$$I_1(x + \Delta x) - I_1(x) = \Delta I_1(x) = [b' I_0 - (b + k) I_1(x)] \, \Delta x. \qquad (5.2)$$

The beam I_2 is similarly affected. Hence

$$\Delta I_2(x) = [b' \, I_0 - (b + k) I_2(x)] \Delta x. \qquad (5.3)$$

However for an homogeneous atmosphere, it is found that I_2 is independent of distance x. Hence

$$\Delta I_2(x) = 0. \qquad (5.4)$$

Eliminating $b' I_0$ between these three equations (5.2, 5.3 and 5.4) gives

$$\Delta I_1(x) = \Delta x(b + k)(I_2 - I_1). \qquad (5.5)$$

Thus subtracting $\Delta I_2(x)$ and normalising

$$\frac{\Delta I_1(x) - \Delta I_2(x)}{I_2(x)} = \frac{-(b + k)(I_2 - I_1) \, \Delta x}{I_2(x)} \qquad (5.6)$$

and from equation (5.1) this can be written

$$\Delta C(x) = -(b + k) \, C(x) \, \Delta x. \qquad (5.7)$$

Thus the contrast decreases with increasing distance. The solution to this equation is of the exponential form, namely

$$C(x) = C(x = 0) \exp \, [-(b + k) \, x]. \qquad (5.8)$$

The sum of the scattering coefficient (b) and the absorption coefficient (k) is called the *attenuation coefficient* or the *extinction coefficient*. (This derivation assumes only single scattering. The inclusion of multiple scatterers, e.g. in dense clouds, leads to more complex expressions than equation (5.8).) Defining a threshold such that the contrast has a value of 2 %, gives the meteorological range L_v

$$L_v = 3.9/(b + k).$$ \hfill (5.9)

Measurements of visibility are undertaken by means of an instrument called a nepholometer.

Taste is usually the last sense to be affected and, under normal ambient conditions, pollutants are not usually detected by taste. For instance the taste threshold of SO_2 is of the order of about 1000 $\mu g\,m^{-3}$ which is an uncommon but by no means impossible (even at present) concentration in industrialised and urban areas.

SUMMARY

The damage caused by pollution in our atmosphere is difficult to assess, especially in terms of human or animal health, since pollution may provide a stress sufficient in itself to cause a reaction (ill health or death) or may be a 'catalyst' in such bodily deterioration. It is thus time-consuming to pinpoint the pollutants of most danger, since their effects may also be slow and/or cumulative and/or synergistic. In this chapter damage has been described in terms of the effects on inert material such as fabrics and building stone—a damage perhaps easier to quantify in terms of cleaning costs; and the damage to plants and commercial animals. Some uses of experimental animals may be justified if the scaling up to human body weight is feasible. The human response has been examined in terms of individual and population response for the affected senses. Legislative limits were introduced as threshold limit values, but it is recognised that different members of a population have their own personal threshold value for each pollutant. Many pollutants today are respirable and the biology of the respiratory tract, together with the body's defence and removal mechanisms have been outlined. Odours and reduced visibility have long been regarded as particularly unpleasant aspects of pollution. Only now, with the lowered ambient levels of smoke and SO_2, can researchers make more thorough quantitative investigations into the possible or actual physically harmful effects of pollutants, whose qualitative nature has been appreciated for some time.

6
Fuels and Industrial Fuel Usage

It has been noted that the majority of air pollutants are generated directly by combustion (or are secondary pollutants created from combustion-generated species). Combustion relates to the process of oxidation of fossil fuels. Hence, in order to understand fully the genesis of contaminants, it is necessary to discuss aspects of coal, oil (and gas) formation in order that the chemical pollutant species can be predicted. The seven major combustion-generated pollutants and the aerodynamic engineering techniques currently used to minimise pollutant emissions are then discussed.

Excepting the use of fuel for space heating (a requirement which is common to industrial, commercial and domestic use), much of the fossil fuel needed by industry is either for steam raising in a boiler when the fuel is burned solely for its heat release (e.g. in electrical power generation) or for a variety of other purposes when the heat is required to maintain chemical processes (e.g. steelmaking, incineration). In this latter case the combustion vessel is known as a furnace. This chapter is concluded by a description of typical boilers and furnaces with special reference to their pollutant emissions.

THE ORIGIN OF FUEL

The fuels of interest for air pollution studies are the fossil fuels: coal, oil (and gas) and their derivatives (e.g. coke). Combustion of these carbon-based fuels results in the formation of the oxides of carbon and hydrogen and of any impurities within the fuel (e.g. sulphur). Typical chemical compositions for these fuels are given in table 6.1.

Both coal and oil are formed by anaerobic decay of organisms under particular environmental conditions. Most of the present coal reserves were formed during the Carboniferous some 300 million years ago when much of the Earth's surface was covered by forests which eventually died and sank into the primeval swamps. Anaerobic decay together with compaction and slight heating (over geological time) during burial led to coal formation. Charcoal is formed from wood by the action of heat alone. Peat is a low grade fuel formed from vegetation which has undergone anaerobic decay without the compaction required for coal formation. Large reserves of peat are found in Russia, Scandinavia, Germany, Canada and the UK. After drying, the water content of peat is still as high as 20–30 %.

Table 6.1 Typical percentage chemical composition of some fossil fuels.

Fuel	Moisture	C	H	O	N	S	Ash
Crude petroleum	0	85–90	10–14	0.06–0.4	0.01–0.9	0.1–7	0
Anthracite	8	78	3	1	1	1	8
Coke	8	82	1	0	2	0	7
Dried fuels							
Wood		50	7	43			
Bituminous coals (medium volatiles)		77	5	6	1	1	10
Sub-bituminous coals		57	6	23	2	1	11
Lignite		41	7	46	0.5	0.5	5

The degree of coalification is known as the *rank*, which increases through the coal series from lignite to high rank anthracitic coals. As the rank increases, so does the carbon content (see table 6.2) whilst the percentage of oxygen and hydrogen both decrease. Bituminous coals are classified as containing over 20–25 % of volatile matter and over half the world's coal reserves are of this type. Anthracite contains less volatile matter (about 10 %) and is found in South Wales (UK) and in Pennsylvania (USA) but seldom elsewhere. As a fuel it is virtually smokeless.

Table 6.2 Composition of fuels ranging from wood to anthracite illustrating progressive changes in % C; % H_2O and % volatiles.

Fuel	% C (of dry ashless weight)	% H_2O	% volatiles (dry ashless weight)
Wood	50	—	75
Peat	60	80–90	65
Brown coal	60–70	40–60	> 50
Lignite	65–75	20–40	40–50
Sub-bituminous coal	75–80	10–20	45
True bituminous coal	75–85	2–10	18–40
Semi-bituminous coal	90–92	1–5	15–20
Semi-anthracitic coal	92–94	< 5	8–15
Anthracite	92–94	1	< 8

Oil deposits were probably formed by anaerobic decay of plants and/or animals in a marine environment. Degraded organic matter which is of an oil-like nature tends to collect in porous rock 'sandwiched' between two impervious layers. The best sites for finding oil are in anticlines where the rock forms a dome in which oil (and gas) collect. Oil exploration is made difficult because it is only oil-bearing rock types that can be identified and not the oil itself.

Whilst coal can be used directly in a combustion process, oil must first be refined. Separation is undertaken initially by means of distillation, absorption or solvent extraction. These techniques separate the different hydrocarbons, which are then refined further for many distinct uses. The chemical structure of the hydrocarbons must often be changed, usually using a catalytic cracker, and impurities, notably sulphur, are often removed before final blending. As oil reserves become depleted, several countries are researching into the techniques of oil manufacture from coal. Perhaps the most successful are the giant Sasol plants in South Africa—petrol from these has been on sale to the private motorist for several years. Other countries (e.g. Brazil) are using alcohol mixtures, deriving their hydrocarbons from vegetable matter such as sugar cane, cassava etc. Table 6.3 gives the coal equivalent for various other fuels.

Table 6.3   Coal equivalents of various fuels.

Fuel type	Amount of fuel	Coal equivalent (tonnes)
Petroleum oil	1.0 tonne	1.7
Natural gas	2.91×10^{10} J (= 8076 kWh = 276 therms)	1.0
Nuclear and hydroelectricity		Amount of coal to produce same amount of electricity at modern coal fired power station efficiencies.

SOLID FUELS

The combustion of coal results in oxidation of some of the carbon (to CO and CO_2) and at the same time the release of much of the volatile fraction, usually at temperatures in the region of 1300 K. However some carbon remains unoxidised ('fixed' carbon) as coke, which can itself be burned. Since smoke emissions are associated with volatile emissions and entrained uncombusted carbon particles, coal combustion is inherently smoky; but coke is 'smokeless'. As a smokeless fuel it was in plentiful supply as a byproduct from town gas production. With the demise of the land based gas industry (in the UK at least) it is no longer readily available, although carbonised coal (coke) is prepared, on site, for the special requirements of the steel industry. Smokeless fuels must now be prepared specifically and a wide range of solid fuels is available. In the UK these are marketed under various brand names (such as Sunbrite, Coalite, Homefire) by the National Coal Board. These are used both by industry and the domestic sector. Room heaters are now capable not only of providing a 'clean' fire but also are able to support a full central heating system.

All coals contain some non-combustible matter which remains as ash. Good quality coals contain about 8 % and the range is generally between 2 and 20 %. Sulphur, as an impurity, may be associated either with the ash or with the volatile fraction; the better quality coal (anthracites) containing less.

Coal deposits form sedimentary bands (see figure 6.1) from which the mineral is abstracted either directly when it outcrops or by mining or quarrying when there is an overburden. Coal burning has been associated with air pollution (or vice versa) ever since its systematic mining at the end of the twelfth century and its widespread use by the mid thirteenth century, after which it became repeatedly a subject for public complaint.

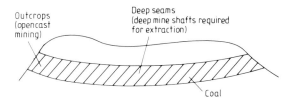

Figure 6.1 Schematic diagram showing positioning of coal in the geological column.

The combustion of coal requires the addition of energy to the fuel–oxygen mixture until ignition (see figure 1.6) after which the exothermic reaction proceeds with the products of complete (CO_2, H_2O) and incomplete (CO) combustion.

The importance of coal as a fuel over the last century is seen in figure 6.2. By the beginning of this century over 70 % of all fuel consumed was coal (oil assuming a greater, if temporary, importance during the second and third quarters of this century). Before 1900 wood burning was predominant, more so in the USA where wood availability was better than in Europe. With the current fashion for a return to the 'simple life' many people in North America and Europe have reintroduced the wood-burning stove on to the domestic scene. In the past few years over 100 000 homes in the UK have had such a stove installed and sales in the USA are currently more than one and a half million per year. Present day stoves are based on a design made in 1740 by Benjamin Franklin. The slow combustion encouraged by such stoves (to minimise restoking) results in the discharge of a large number of hydrocarbons. As a pollution source they rank high, despite the negligible sulphur content of most woods. The constituents of the wood smoke are similar to those of cigarette smoke (see Chapter 1) and include many known carcinogens. Although the number of wood stoves which will be in use in the UK will be limited by the availability (and price) of wood, the self-limiting process may not occur in the USA before domestic smoke once again becomes a pollution problem—it is likely that the USEPA will have to consider other methods to limit their use.

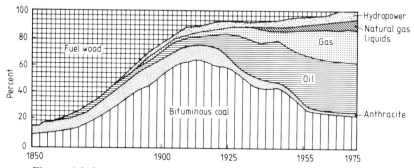

Figure 6.2 Sources of energy in the United States. After Luten (1974).

LIQUID FUELS

Petroleum-based liquid fuels cover a wide range of fractions for different applications, e.g. combustion in a boiler or furnace, petrol, kerosene etc. Combustion of these fuels also results in the formation of carbon oxides and water.

Each hydrocarbon can be characterised by its boiling point (important for vapourisation or evaporative losses), its viscosity (important for pipeline transport) as well as chemical composition. The kinematic viscosity is expressed in $m^2 s^{-1}$ (or centistokes where 1 centistoke equals $0.1 \ m^2 s^{-1}$) at a given temperature or (in some countries, for example the UK) as a Redwood number. This gives the time for a standard volume of the oil at 311 K ($38 \ ^\circ C$) to pass through a standard orifice and is thus measured in seconds—the larger values being associated with highly viscous oils.

FUEL COMBUSTION

For industrial combustion the fuel may be in the solid phase, pulverised or gasified. Liquid fuels are often atomised. The stages of combustion are depicted in figure 6.3 and these steps must be borne in mind when discussing the combustion chamber and burners. Chemical reaction rates are relatively high and this means that the control of the rate of mixing is important. High temperatures must be maintained throughout the combustion changes, otherwise high temperature (2000–3000 K) flame reactions may not be mimicked by the low temperature reactions in the bulk of the gas. These important factors can be best remembered as 'the three Ts' of time, temperature and turbulence. Good combustion needs a sufficient contact time, high enough temperature and good mixing-induced turbulence.

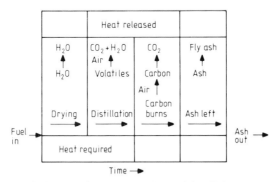

Figure 6.3 Schematic representation of fossil fuel combustion.

There are seven major combustion generated pollutants:

(*a*) CO
(*b*) NO_x
(*c*) unburnt or incompletely oxidised hydrocarbons
(*d*) smoke and solids
(*e*) SO_x
(*f*) lead compounds
(*g*) hydrogen chloride.

Carbon monoxide
Carbon monoxide, a toxic gas, may be formed intentionally (e.g. in some metallurgical processes which demand an excess concentration of CO) or inadvertently. As a product of incomplete combustion it may be generated as a result of several types of mismanagement of the combustion process such as incorrect (real-time) control, badly designed equipment or operation outside design limits. It can be a serious problem in spark ignition engines especially when idling and decelerating (see Chapter 7), but is consumed by oxidation with free hydroxyl radicals:

$$CO + OH \rightleftharpoons CO_2 + H.$$

This is a fast reaction over a large temperature range.

Nitrogen oxides

Again, partially oxidised nitrogen is a result of incomplete combustion and hence not a stable end product. Nitric oxide (or nitrogen monoxide NO) arises from oxidation of N_2 which is an 'unnecessary' component of the air usually used for combustion of the fuel, and from nitrogen present in fossil fuels (but not petrol). Nitric oxide is formed at high temperatures and pressures and can be oxidised further (usually after emission) to nitrogen dioxide (NO_2)

$$N_2 + O_2 \xrightleftharpoons{2000 \text{ K}} 2NO \xrightleftharpoons{8000 \text{ K}} 2N + 2O.$$

NO formation from nitrogen in fuels occurs rapidly and is unaffected by combustion condition changes whereas formation from N in the air is highly dependent upon the combustion process. It is possible that large concentrations of both CO and NO (formed on the fuel-lean side of stoichiometric combustion†) may react and lead to mutual destruction—a slow reaction which may be assisted by a catalyst

$$2NO + 2CO \longrightarrow N_2 + 2CO_2.$$

Hydrocarbons and smoke

In combustion processes, the partial oxidation which results in the production of carbon monoxide can also lead to the emission of unburnt fuel (as a consequence of the insufficient oxygen). In an internal combustion engine this provides the major source of pollution, especially in engines of low efficiency. The survival of unburnt fuel is often the result of wall effects, possibly a symptom of poor mixing or combustion chamber design associated with the boundaries.

Smoke and particulates are formed under similar circumstances and in the context of combustion chambers can, in general, be eliminated.

Sulphur oxides

SO_2 is a result of the sulphur present in the fuel (see table 6.4). Most comes from coal combustion—the sulphur content of oil may be less, although many of the residual fuel oils used have a sulphur content of about 4 %. (Since this has a high calorific value it is equal (in SO_2 emissions) to a 2.5 % S content coal.) With the increasing consumption of lower grade fuels (both coal and oil), the sulphur content is seen to be rising, which would of course increase the total SO_2 emissions (all other factors remaining constant).

SO_3 may be emitted directly (primary pollutant) or formed by oxidation of SO_2 in the atmosphere (secondary pollutant)—see Chapter 1 for details of these reactions.

---

†A stoichiometric combustion process is one in which exactly all the oxygen is used, i.e. the chemical equation describes precisely the process occurring.

Table 6.4   Sulphur content of fuels.

Fuel	S(%)
Bituminous coals	1.5–2.5
Coal (eastern US)	up to 6
Coal (western US)	0.5–6.0
Coke	1.5–2.5
Petrol	0.1
Kerosene	0.1
Diesel fuel	0.3–0.9 } rising
Fuel oils	0.5–4.0 }
Natural gas (e.g. North Sea)	zero (unless added)

Lead and HCl

Lead emission is largely a result of petrol combustion. Tetraethyl and tetramethyl lead (TEL, TML) are added as anti-knock agents to petrol at concentrations of approximately 400 ppm. They are emitted (in the exhaust gas) largely in the form of PbBrCl (see Chapter 1).

The existence of organic chlorine compounds (e.g. in PVC in incinerator fuels) frequently leads to the emission of hydrogen chloride gas. HCl is also emitted in coal combustion especially if the coal contains traces of NaCl.

The above pollutants are all primary pollutants—secondary pollutants such as nitrogen dioxide, sulphuric acid and photochemical smog constituents may form directly as a result of combustion generated pollution.

Effect of design and operation on emissions

In engineering terms the production and emission of pollutants can be identified as a failure to burn the carbon and hydrogen in the fuel as a consequence of poor mixing, inadequate residence times, insufficient air and incorrect or inhomogeneous temperature fields. In many industrial combustion processes control is undertaken at the mixing stage. Poor mixing can lead not only to a combustion process which is only partially complete but can also result in a large volume of unused oxygen which carries away excess heat. Hence in poorly controlled combustion there are large heat losses in the exhaust gases and a higher probability of oxidation of sulphur (and nitrogen in the air). *Good aerodynamic design* is therefore vital in order to minimise excess air. Other real-time engineering controls are required for maintaining the correct residence time and controlling heat extraction, taking care to ensure that the combustor is not overloaded nor the exhaust gases quenched too early.

The major variables (residence time, excess air and temperature) are frequently interrelated. In figure 6.4 this relationship is shown for stoichiometric ratios greater than unity (namely excess air) for a small oil-fired combustor.

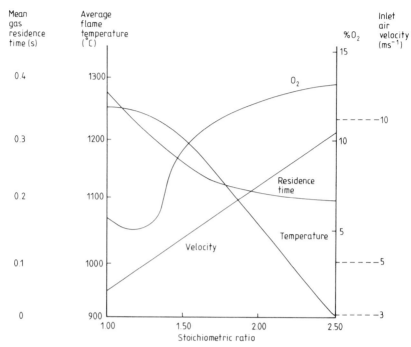

Figure 6.4 Inlet air velocity, flame temperature, exhaust gas oxygen concentration and residence time as a function of stoichiometric ratio in a small oil-fired installation. Data courtesy SERC.

Optimisation of combustion is governed by the rapidity of combustion of CO to CO_2. Soot is formed by processes as yet not completely understood but is known to be composed primarily ($> 90\%$ $^w/_w$) of spherical carbon particles about 1×10^{-8} to 5×10^{-8} m in diameter (0.01 to 0.05 μm). Soot particles can be oxidised, but this is a slow process. The use of pulverised bituminous coal or heavy fuel oil results in the formation of hollow carbonaceous spheres, the presence of which leads to a critical design problem.

At high temperatures, fixation of atmospheric nitrogen can be a problem since this occurs under conditions normally associated with complete combustion of hydrocarbon-based fuel (namely excess air and high temperatures). In fact nitric oxide concentrations are often higher than calculated. Some typical NO yields as a function of combustion chamber wall temperature are given in table 6.5.

Studies of sulphur oxidation were initiated originally because of corrosion problems rather than a concern for air pollution—at low concentrations (e.g. 5 ppm) SO_3 presents problems as a corrosive agent and SO_2 is itself an acidic gas. Although most sulphurous impurities in fuels are oxidised to SO_2 about 5 % of this is further oxidised to SO_3 (and thence by solution in water to sulphuric acid). Consequently in, for instance, a domestic water heater it is essential to use a low sulphur content fuel.

Table 6.5 Yields of NO as a function of furnace wall temperature.

Furnace wall temperature (K)	Yield of NO(%)
1810	0.26
1920	0.41
2030	0.77
2140	1.30
2250	1.55
2360	1.75

Table 6.6 Size ranges of particulates removed by different types of control equipment.

Control apparatus	Particle size range (μm)
Cyclone	> 10
Cloth filters	100−1
Paper filters	100−0.1
Wet scrubbers	30−0.1
Electrostatic precipitators	30−0.1

Table 6.7 Fractional efficiencies for various particle sizes for a multicylone.

Particle size (μm)	Fractional efficiency (%)
0−2.5	23.4
2.5−5.0	56.4
5.0−7.5	70.5
7.5−10.0	80.1
10.0−15.0	90.1
15.0−20.0	94.5
> 20.0	98.5

The non-volatile fraction of the fuel is left as ash which either remains in the grate or is contained in the waste gases (fly ash). If the latter is the case control procedures such as use of electrostatic precipitators should be followed (see Chapter 8). The ranges of particle sizes for which the different control procedures are effective are given in table 6.6. The efficiency of removal is a function not only of the apparatus used but also of the particle size, and data are often given in terms of the 'fraction efficiency'. Table 6.7 gives some observations of the removal of fly ash by a multicyclone employed by a power station boiler. Once in the atmosphere smaller particles, of course, remain suspended longer and large particulates are no worse as air pollutants than locally emitted particulates (dust deposits near the source).

BOILERS

There are two basic types of steam-raising boiler: the shell boiler and the water tube boiler. The shell boiler is relatively inexpensive with internal fire boxes. It may be of a vertical or horizontal design or may utilise waste heat rather than primary fuel. Conversely the advantages of the water tube boiler are its high efficiency and higher operating pressures. The final choice will depend upon several factors: the amount and pressure of steam required, whether the demand will be continuous or intermittent, the quality of the incoming water and the physical space available to house the boiler.

Many of the older boilers still in existence are of the horizontal type. A type common at one time in the UK was the Lancashire boiler (figure 6.5). One of the major advantages of this was its tolerance to poor handling and its ability to use as fuel coal, sawdust and other combustible materials; although not being readily convertible to oil. These boilers had a long working life, often more than forty years, much greater than many other otherwise equivalent designs. The dimensions were sufficient for human entry into the boiler for the purpose of cleaning. Usually there were two independent furnace tubes. Fuel was fed in, originally by hand, on to the grate towards the front of the boiler. Primary air was drawn in (and often a secondary air supply). The heat was led away along the length of the furnace tubes which were immersed in water. Heat exchanges across the walls resulted in the creation of steam which was then directed away for use elsewhere. Often some recirculation of the waste gases below the boiler was used, perhaps to preheat the incoming air. This recirculation could save 1 % of the fuel. Simple in conception and construction, the Lancashire boiler was at one time very much in evidence in large numbers. However its size and its inability to be converted to oil has led to its decline. The principles are however similar for the other boilers described (in less detail) below.

One of the most frequently commissioned horizontal boilers nowadays is the *economic boiler* (figure 6.6). The initial stages are similar to the Lancashire boiler, but gases are recirculated to a greater extent, so that the overall length is reduced from about 8.5 m to only 5 m. Its useful life is generally estimated at about fifteen years. The water used in the boiler must be of high quality and free from dissolved solids. Indeed it may be necessary to pretreat the water with an ion-exchange apparatus.

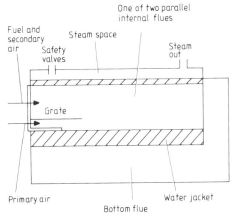

Figure 6.5 A Lancashire boiler.

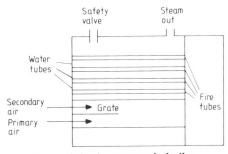

Figure 6.6 An economic boiler.

The *thermal storage boiler* (figure 6.7) is designed to cater for fluctuating demands. A reservoir of water is maintained which acts as a 'thermal flywheel'. When demand is low, the water is at its highest level and the large volume of water retains the heat which is still being supplied at a constant rate. This stored heat is then held in reserve until peak demand is reached when the heat is released, the water evaporates and the level drops. With this operation demands of 25–40 % above and below the mean can be satisfied. Regulation is by means of two regulators.

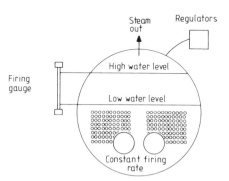

Figure 6.7 A thermal storage boiler.

Variations of these boiler types exist. For example, the Cochran boiler is of vertical design, thus minimising land usage. Two other subclasses of shell boilers are *welded steel boilers* and *sectional boilers*, both of which may be used as hot water or steam boilers.

Water tube boilers (figure 6.8) can be more expensive, yet are essential for large outputs, especially at high pressures, as for example in power stations where pressures in excess of $2 \times 10^6 \, \mathrm{N\,m^{-2}}$ (20 atmospheres) may be required. They can accept almost any fuel, require water free from dissolved solids and can handle fluctuations in demand of $\pm 20 \%$. However they require more frequent attention than shell boilers. Figure 6.8 shows the basic parts. Fuel is usually added by a mechanical stoking device (see later in this chapter). The hot gases rise and warm the outside of a set of tubes through which the water is circulating by means of free convection. Some water tube boilers may utilise several steam and water drums and possibly forced convection.

A further classification is the *magazine boiler* which is specially designed to burn low volatile solid fuel. Control of the air supply permits a rapid transition from 'slumbering' to full output (up to 265 kW).

On occasions it is more expedient to use hot water (rather than steam) to transport the heat for space heating. In such a system the pipe circuitry is considerably simpler since the circulating water requires less volume per unit of energy carried than does steam. Again several designs exist.

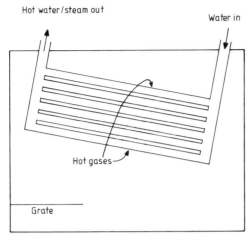

Figure 6.8 A water tube boiler.

Boiler instrumentation

Optimisation of boiler operation requires careful surveillance. The aim is to get the most steam for the least amount of fuel whilst avoiding a burst boiler. The basic instruments required to ensure that the process is at all times working satisfactorily are a steam pressure gauge and safety valve. The gauge glass shows the water level and a low water alarm is installed. Temperature control is important and commonly measured by the use of thermocouples in the flue. Monitoring of carbon dioxide is a common method of ensuring that complete combustion is maintained at all times.

Pollutants

The generation of pollutants in an industrial process depends upon the fuel characteristics, the temperature in the combustion chamber, the mixing and the residence time. Thus a good aerodynamic design is required. In large boilers (such as at power stations) the surface to volume ratio is reduced and a high degree of control is required to ensure there are no regions of poor mixing and subsequently incomplete combustion.

Pulverised fuel combustion depends in addition upon the volatile component, the fineness of the fuel and the excess air. Oil firing also requires excess air but usually less than 2 %, and the fact that this is such a small excess makes it more difficult to control the combustion process. The emission of particulates depends upon the quality of atomisation (i.e. mean drop size and proportion of the mass of drops with a diameter greater than about 200 μm).

Recent research on smoke suppression from oil-fired boilers indicates that by using an oil–water emulsion (20 % H_2O to 80 % oil) better results can be attained (any larger amounts of water will probably make the process uneconomical). It has been shown that the burner design may be critical here, especially with respect to the degree of 'swirl', which enhances mixing and helps to maintain the required residence time. Carbon monoxide emissions from large boilers are usually low (of the order of 0.01 %). Continuous infrared monitoring with a control feedback can maintain the correct air–fuel ratio. Burner conformity is also important in furnaces with large numbers of burners. If the oxygen supply is unevenly distributed, excess CO emissions will result from a part of the furnace. To minimise this, the *total* air supply must be increased, but this will result in excess NO production in the remaining portion of the furnace. This example illustrates the problem of NO_x formation. Nitric oxide (NO) is emitted when air is used for combustion at long residence times and high temperatures (thermal NO formation) or by preferential NO formation from the nitrogen in the fuel under fuel-lean conditions (fuel NO formation).

Sulphur is present in most fuels. In the combustion process it is oxidised to SO_2 (or SO_3). Using coal as a fuel, a typical untreated emission might contain 800–4000 ppm. When it is remembered that at ground level the requirement may be of the order of 0.1 ppm averaged over a 24 hour period and/or 0.02 ppm as an annual average (US figures) it is obvious that control at source may be a worthwhile project environmentally. However, since SO_2 reduction is likely to be by between one and two orders of magnitude only, adequate atmospheric dispersion or dilution is also vital. Reduction of SO_2 at source may be accomplished by

(*a*) desulphurisation of fuel;

(*b*) desulphurisation of waste gases;

(*c*) use of sulphur absorbing additives;

(*d*) use of new techniques e.g. coal gasification/gas desulphurisation/ sulphur retention in chamber as in fluidised bed combustion.

Some of these techniques are worthy of more detailed discussion.

Control methods
Since SO_x (i.e. SO_2, SO_3) and NO_x are formed under conditions of excess air, one control method is to restrict the volume of excess air used (to say less than 2 %). This should restrict NO_x formation especially, for so long as the flame temperature does not rise substantially. However accurate calculation and maintenance of the air–fuel ratio is required. The use of a low volume of excess air will often result in a higher emission of combustibles.

Staged combustion with heat removal between stages helps to decrease flame temperatures and hence NO_x formation. However its operation is based on empiricism and is at present a comparatively unsophisticated technique.

Flue gas recirculation can be applied to many industrial and domestic situations, including power station boilers. The peak combustion is reduced, minimising both NO_x emissions and the formation of soot.

However, the flame stability may be adversely affected. There may be the engineering problems of large lengths of ducting and pumping for large volumes of gases.

Gasification of coal and residual fuel oil is used in power generation. In *fluidised bed combustion* the material forming the bed is kept in suspension by air jets such that it behaves as a fluid. Many pollutants (especially sulphur compounds) remain trapped in the bed when materials such as limestone or chlorates are added and thus emissions are reduced directly. Lower combustion temperatures again mean low NO_x. This technique is still not fully developed and problems were initially encountered in scaling up pilot plants to power station requirements. It is intended that fluidised bed combustion be used with vertical shell boilers (≤ 4 MW), with horizontal shell boilers (> 4 MW) and with water tube boilers ($\gg 4$ MW). At present, there are boiler installations in the UK operating (at atmospheric pressure) at up to 30 MW. In the USA all but one of the locations (Rivesville Power Station, West Virginia, 100 MW) are also in this range, whereas in West Germany an advanced plant at Volkingen is currently operating at 667 MW.

The type of firing can be crucial in the formation of pollutants. However this discussion is reserved until after the description of furnaces.

FURNACES

A basic distinction between furnace types is between *slag tap* and *dry bottom* furnaces. The former type is designed to cause molten ash particles to accumulate on the walls. This liquefied ash then runs down the furnace walls and out of the furnace at the bottom through the slag tap. In this type of operation, ash is not therefore an air pollution problem. On the other hand a dry bottom furnace is designed to cool all residual ash particles below their melting point before they strike the walls or other heat absorbing surfaces. With this design about 80 % of the ash leaves the furnace as fly ash, compared to 50 % for the slag tap furnace. Often the collected ash is reintroduced in the lower part of the furnace which may be modified to encourage high flame temperatures in the lower part of the furnace so that the collected ash is converted to a coarse slag. A more efficient refinement exists in the *cyclone furnace* in which crushed coal is fired into a water-cooled, refractory-lined cylindrical chamber, discharging gases nearly horizontally into a water tube boiler. The combustion is so intense that a small proportion of molten ash coating on the walls is vapourised resulting in about 85 % of the ash being retained as a molten slag. Thus emissions to the atmosphere are less. However, this dust is extremely fine, difficult to collect and an obvious respiratory danger if emitted.

An alternative, and complementary, subdivision of furnaces is based on the degree of contact between fuel and industrial material being worked. There are three distinct groups:

(*a*) Furnaces in which the material is separated from the fuel and gases. Typical examples are (boilers), coke ovens, oil-fired furnaces and oil refineries. The production of coke from coal is still of importance for on-site use in the steel industry although in previous years this process was far more widespread since, in the UK, town gas production (before the advent of North Sea gas) relied on the carbonisation of coal to coke, releasing hydrogen, carbon monoxide and methane as a fuel for distribution (this process is still used to produce town gas in parts of Ireland). In this process the waste product was coke which soon became much in demand as a smokeless fuel as bans on domestic use of the inherently smoky coal have been introduced. To produce coke, coal was heated in closed horizontal retorts (figure 6.9) by means of producer gas. Horizontal retorts were superseded by vertical retorts which, unlike the horizontal beds which had to be operated on a batch process, could be used continuously. The white hot coke discharged from the retorts was then cooled, using large amounts of water, and tipped. Pollutant emissions can occur at time of discharge and recharge (especially coal dust as the retort is charged) as well as when the coke is quenched.

Figure 6.9 Horizontal retorts once used for coke and town gas production.

(*b*) Furnaces in which the material and fuel are in contact. This class includes blast furnaces and cupolas. The most frequent use made of blast furnaces is in the production of iron (although smaller blast furnaces are also used in lead smelting). In the furnace the fuel must be burned in such a way that the gases evolved are available to enter into the chemical reaction that is needed in the process. A typical blast furnace as shown in figure 6.10 may be over 20 metres high. Emissions contain carbon monoxide, dust and smoke although these can be controlled effectively with modern-day technology. Perhaps the biggest problem is when the furnace load 'slips' and a sudden release of gas is sufficient to operate the safety valve on the top of the furnace. If this release, containing high volumes of dust, is allowed to escape into the atmosphere, a local pollution problem is inevitable. Slipping can be controlled—largely by inducing minor slips of insufficient magnitude to cause a problem of emergency venting. Cupolas can also be responsible for heavy smoke emissions. They are a smaller type of shaft furnace used largely for melting of iron and steel, often including a proportion of scrap metal.

Figure 6.10 A blast furnace. British Steel's Scunthorpe works. Photograph courtesy British Steel Corporation.

(*c*) Furnaces in which the material is in contact with flames of the gaseous products of combustion. Typical furnaces in this class are hearth furnaces, pottery, brick, lime and cement kilns. Many of these furnaces are to be found in the clay industries. Kilns may be fired continuously or intermittently, automatically or by hand, by coal or oil burning. Although flue gases are usually passed through some type of control stage (often using cyclones or electrostatic precipitators), dust emissions still occur in addition to possible production and emission of smoke and waste gas associated with any combustion process.

It is not the intention to describe each of these industrial furnaces in detail as they are well discussed in other texts (see e.g. Parker 1978, Meetham *et al* 1981).

TYPES OF FIRING

Over the last few decades there has been a transition from manual firing to automatic firing in furnaces and boilers. Hand firing was notoriously smoke-producing. The volatile content was a useful index of smoke-forming tendency; a problem which has been virtually eliminated by the advent of automatic firing together with smoke abatement policies.

Figure 6.11 depicts a basic *overfeed* fuel bed ideally suited to hand firing. The air flow is upwards through the fuel which is held on a fixed bed or grate. The fuel flows downwards as it is combusted, with the ash falling into the collector beneath the bed. New fuel was spread on top of the burning fuel bed, and to mix the fuel successfully considerable muscular strength and effort was necessary. In this and many other bed configurations, the new fuel encounters a gas flow with minimum excess air. Volatilisation is encouraged in the upper parts of the bed. Pollutants created cannot be further combusted in the upper part of the bed unless *overfire jets* are used. This idea has been in use for over 100 years and creates a good availability of excess air throughout the combustion chamber. The flow itself can also enhance the mixing within the burning fuel.

The problems of the overfeed bed can be obviated to some degree by the use of an *underfeed bed* in which the fresh fuel is added to the burning fuel by an *underfeed stoker*—an automatic device for introducing new fuel. In an underfeed bed, new fuel is added to the hottest part of the combustion zone and thus there is ample air for complete combustion. Some fly ash may be produced necessitating the introduction of dust collection devices into the waste gas stream. This practice is however highly efficient in the control of smoke emissions. Underfeed stokers are suitable for use with sectional boilers and can be used for horizontal and vertical shell boilers and small water tube boilers.

Further examples of stoking mechanisms are all designed to spread and mix the fuel into the burning zone. A *spreader stoker* is either a mechanical device or a system by which jets of air or steam throw solid fuel into the furnace where it falls on to the grate, which may be stationary or travelling. Essentially this is a form of overfeed stoker and is thus likely to be smoky. In fact some burning of suspended particles occurs. This can be encouraged again by the use of overfire jets in an attempt to cut down on particulate emissions. Spreader stokers are suitable for water tube boilers with a rating of less than 80 MW.

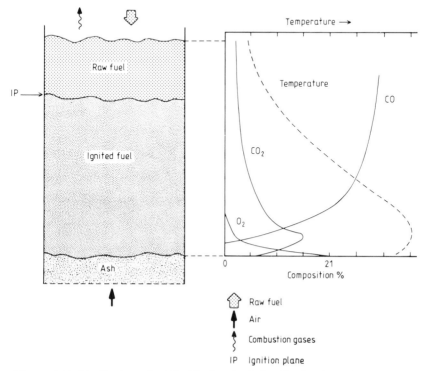

Figure 6.11 Idealised overfeed fuel bed and relative distribution of temperature and products of combustion. Courtesy US Bureau of Mines and US Government Printing Office.

Travelling grate stokers utilise the motion of the grate itself to encourage mixing. Some types vibrate and this may cause an air pollution problem by dislodging more particles and creating a dust nuisance. Although they have a high capital cost, they have a long lifetime and can be used in units with ratings from 90 kW to 70 MW.

The fuel itself may be pretreated in order to minimise emissions and optimise the combustion. Pulverisation of fuel (coal) to a size of about 50 μm leads to very rapid combustion. The cost of the fuel preparation is relatively high and thus such plants tend to be uneconomic on a small scale. There are sometimes problems of fly ash accumulation (and hence disposal)—between 50 and 80 % of all ash in coal leaves the combustion chamber as fly ash and hence dust control equipment (see Chapter 8) must be installed. However this dust does tend to be of uniform size and thus can be used as a cement additive in road building. During the combustion the particles of fuel become heated and plastic. The rapidly increasing temperature can cause volatilisation of gases within the particle causing the particle to swell to two to five times its original size. The gas is released and the fragmented particle is carried away as ash.

Grates

The basic types of grates are fixed and moving; although some furnaces may be better described as grateless. In order to encourage mixing a fixed grate may be of a stepped design (figure 6.12). However, there is a larger variety of designs of the travelling or moving grate type. These include chain grate, roll grate, forward feed grate, backward feed grate, agitated grate, rotary grate, rotating plate grate and combustion cone. A chain grate is on a kind of moving conveyor which gradually moves the fuel through the combustion zone. It is therefore a continuous process in contrast to the essentially batch process of hand stoking to a fixed grate. Care must be taken to ensure that material does not clog the sprockets at the edges of the grate. The chain grate itself may provide a horizontal bed but a more efficient method is to combine a chain grate with an inclined grate to enhance mixing by gravity.

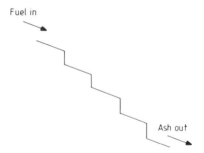

Figure 6.12 A stepped grate.

A roll grate is a series of rolls at different vertical levels each of which carry a chain grate. In this the fuel is turned several times. Figure 6.13 shows a forward feed grate in which the fuel is pushed forward into the combustion zone by a continuous motion. Rotary grates are fed by fuel at the centre which is then turned by the system and gradually moves to the periphery as combustion is completed. A similar idea is used for a rotating plate grate.

Grateless furnaces include shaft ovens (a vertical chamber with some agitation), horizontal drums (which rotate about an axis at a few degrees to the horizontal) and fluidised cyclones (where the fuel is fed in tangentially at the top to meet a stream of secondary air rising from the bottom of the furnace). The material burns in a fluidised state with high efficiency.

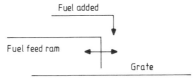

Figure 6.13 A forward feed grate.

Many of the systems described above are to be found in one specialised form of furnace—the *incinerator*. This can be of importance from an air pollution point of view because of the heterogeneous nature of the fuel which may range from hydrocarbons to vegetable and animal stuffs to plastics and PVC. Indeed the composition changes not only from incinerator to incinerator but also at different times of the day, different days in the week and by season. Incinerators used to be sited at great distances from centres of population (which in itself provides a transport problem for the incinerator fuel). Increasingly efficient operation permits sitings in centres of cities, although effective control equipment is vital (see Chapter 8).

SUMMARY

The world has seen rapid changes in the ratios of the major fossil fuels used directly (coal, gas, oil) and indirectly (via electrical power generation). The direction of global energy policies in the future is as yet undecided but is likely to include more diverse power sources including existing fuels, synfuels (e.g. from shales and coal) and more use of renewable energy sources.

The seven major combustion generated pollutants have been identified in this chapter as CO, NO_x, HC, smoke, SO_x, lead and HCl, and their chemistry has been discussed.

Boilers and furnaces rely on the combustion of fossil fuels and hence provide a large percentage of the sources of air pollutants. Some details of construction and management are needed to evaluate better methods of control. Many of these techniques, including design of grates and stokers, are currently being researched and developed. (The implementation of control technology for the post-combustion chamber is reserved for Chapter 8.)

7

Mobile Sources

Pollutant emissions from stationary sources are a function of time and efflux characteristics only and dispersion is a function of prevalent meteorological conditions. In the case of a non-stationary source an additional variable is introduced, that of a moving emission, so that concentration calculations are made more difficult. For emissions from traffic moving along a (straight) road, a line source model can be used to calculate concentrations at distances away from the road. (A simple modification may be made to the Gaussian plume model or the numerical models described in Chapter 3.)

Figure 7.1 City-centre traffic can be a major source of urban pollutants. Pollutants become most visible early on cold mornings, especially from diesel vehicles. This picture was taken in Glasgow, Scotland.

The major mobile source is undoubtedly road vehicles (figure 7.1). These can be further subdivided into, for instance, private and commercial vehicles. However for air pollutant considerations a better division is based on the engine type: the two most important of these being the Otto cycle (petrol) engine and the diesel engine. These categories do not completely coincide with the private/commercial classification although a good correlation does exist. Table 7.1 gives a list of engine types, fuels and associated pollutants—which will be discussed in more detail in subsequent paragraphs.

Table 7.1 Vehicle types and pollutant emissions for various engine/fuel combinations.

Engine type	Fuel type	Vehicle type	Major emissions
Otto cycle	Petrol	Cars (also buses, lorries/trucks, aircraft, motorcycles, tractors)	HC,CO,NO_x,Pb
Diesel	Diesel oil	Lorries, trucks, buses, trains, ships, tractors (also cars)	NO_x,SO_x, soot, particulates
Two-stroke cycle	Petrol	Motorcycles, outboard motors	HC,CO,NO_x, particulates
Gas turbine (jet)	Turbine	Aircraft, marine (also rail)	NO_x, particulates
Steam	Oil, coal	Marine	NO_x,SO_x, particulates

PETROL ENGINES

Vehicular emissions have been said to be a prime indicator of an industrialised/civilised (?) population. The extent to which this may be regarded as true is illustrated by car ownership figures for both the UK and the USA which show an increase in population together with an increasing number of vehicles per person. However over recent years, car designs and sizes have changed with the result that more effective emission control (and the elimination of lead from petrol in the USA) make the assessment of pollutant emission trends difficult.

The chemical species here are those associated with the fuel (i.e. vapourised hydrocarbons) and with the exhaust gases (especially CO, NO_x, unburned hydrocarbons). The composition of the waste gases can be determined in terms of $\%^v/_v$; $\%^w/_w$ or mass emissions per unit distance travelled. Figure 7.2 shows the most important points of origin of these pollutants in a well-maintained car. Petrol is a readily volatilised fuel and in the fuel tank the pressure build up which would result from this evaporation is obviated by introduction of a 'breather' vent or pipe into the tank. This still permits evaporation of the fuel. Evaporation of the raw fuel also occurs whilst it is in the carburettor at all times except when running at high speed. Some unburnt fuel (mixed with air) plus escaping exhaust gases from around ill-fitting pistons leaves the car as crankcase blowby—a further hydrocarbon loss. All these losses can be minimised, and even eliminated, by incorprating a recirculation system at the points of emission. In the USA crankcase emissions have been controlled since 1963 (California), 1968 (Federal) and evaporative losses since 1971.

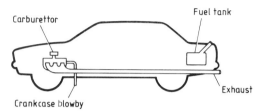

Figure 7.2 Schematic representation of the sources of pollutants in vehicles.

Exhaust emissions are more variable in nature and hence more difficult to control. The composition depends on several variables e.g. air–fuel ratio (figure 7.3), speed and engine condition. Driving conditions play a major role with exhaust emissions high in CO and HC at low and idling speeds and NO_x high at high engine speeds. At low speeds, especially when cold and the fuel mixture is fuel-rich,incomplete combustion is common resulting in the formation of more carbon monoxide. Similarly unburnt hydrocarbons can be part of the exhaust—up to 4 % in traffic holdups.

When cruising, combustion tends to be complete. With excess air, nitrogen can be oxidised to NO and possibly NO_2. Maximum formation occurs with an air to fuel ratio of 16 : 1. These emissions are particularly important in regions of intense solar radiation, which can catalyse reactions leading to the formation of photochemical smogs (see later in this chapter).

It is worth noting that whilst cruising at high speeds, the exhaust flow is also high, whilst at low speeds the exhaust flow is low, although emission of partially oxidised compounds is higher. This can lead to the seemingly anomalous statement that highest emissions occur in deceleration on a volumetric basis—low air–fuel ratio, low exhaust flow; whilst highest emissions on a mass basis occur whilst accelerating or cruising—high air–fuel ratio, high exhaust flow. It is thus important to specify the units for vehicular emissions. In table 7.2 emissions are judged on mass per unit distance (averaged over all operating conditions). These figures illustrate the possible limitation of pollutant emissions by judicious control. For exhaust gases this is perhaps best accomplished by either recirculation or the use of an afterburner. This secondary combustion chamber can be of a direct flame type, operating at temperatures of about 1100 K, or use a catalyst at a lower temperature (600–800 K).

Recirculation can also be advantageous in that this limits the temperature inside the engine, thus reducing the amount of nitrogen oxide formation, perhaps by up to 90 %. However, on the deficit side, the lowered operation temperature usually results in loss of power.

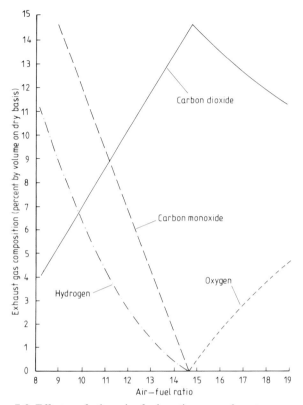

Figure 7.3 Effects of the air–fuel ratio on exhaust composition. Redrawn from Stern A C 1962 *Air Pollution* vol 2 (London: Academic), by permission.

Table 7.2 US vehicle emissions (petrol engines) for 1960 and 1976 in g km^{-1}.

Source	HC		CO		NO$_x$	
	1960	1976	1960	1976	1960	1976
Fuel tank and carburettor	1.74	0				
Crankcase blowby	1.99	0				
Exhaust	6.21	0.10	31.07	2.92	3.29	0.25

Exhaust cooling can also have advantages. The water vapour content of the gases condenses and acts as a trap for other soluble compounds, notably SO$_2$, NO$_x$ and some HC and lead compounds. With this technique the condensed water (which represents the equivalent of about 46 % by volume of the petrol used) typically contains the concentrations of pollutants shown in table 7.3. Much of the lead content originates from the tetraethyl and tetramethyl lead used as an anti-knock agent in petrol. Typical emissions of 0.8 mg g^{-1} of fuel contain particulate lead compounds, 70 % of which by number are in the size range 0.02–0.06 μm and 5 %, by weight, less than 1 μm in size. In the USA most cars built in the last few years have been modified to burn only unleaded gasoline (petrol)—a much needed move towards eliminating a potential hazard from the atmospheric environment. A recent (March 1982) development by Associated Octel in the form of a lead 'filter' to adsorb lead compounds is claimed to remove, on average, 60 % of lead from emissions (50–60 % on motorways/freeways and 80–90 % in urban areas).

Table 7.3 Pollutant concentrations in condensed exhaust gases from a petrol engine

Pollutant	Concentration
Solids	0.24 %
Sulphate ions	72 ppm
Chloride ions	123 ppm
Nitrate ions	0.25 ppm
Lead	15 ppm
HC	60 ppm

Table 7.4   Comparative emissions from petrol and diesel engines.

	Idling	Acceleration	Cruising	Deceleration
Petrol				
CO (%)	6.9	2.9	2.7	3.9
HC (ppm)	5300	1600	1000	10 000
NO$_x$ (ppm)	30	1020	650	20
Aldehydes (ppm)	30	20	10	290
Diesel				
CO (%)	0	0.1	0	0
HC (ppm)	400	200	100	300
NO$_x$(ppm)	60	850	240	30
Aldehydes (ppm)	10	20	10	30

DIESEL ENGINES

In the diesel cycle engine, the air flow is constant independent of engine speed, which is regulated by fuel flow only. Well maintained diesel engines emit fewer pollutants than petrol engines. The diesel 'smoke' often observed is largely due to overloaded engines or poor maintenance. Under such conditions at critical fuel–air ratios the fuel is no longer burnt but 'cracked' and the emissions may contain a variety of hydrocarbons (including some known carcinogens such as benzo(a)pyrene). Some sulphur dioxide may be formed under many operating conditions as a result of the sulphur impurity in the fuel (about 0.5–1 %). Table 7.4 compares the emissions from both petrol and diesel engines under a variety of operating conditions and is based on UK figures. USEPA measurements suggest that the particulate emissions in the USA are excessive. Their regulations require diesel cars to emit no more than 0.6 g mile^{-1} by 1982 and 0.2 g mile^{-1} after 1985. This compares with petrol engine emissions (with catalytic converter) of only 0.008 g mile^{-1}. They anticipate that 10 % of all new cars sold in 1985 will have diesel engines and perhaps 25 % by 1990.

GAS TURBINES AND JET ENGINES

The major pollutant problems here are unburned hydrocarbons, unburned carbon (visible exhaust) and NO_x. The first two can be minimised by design control and the latter by reduction of operating temperature.

Ships in port can be responsible for emissions (see figure 7.4), especially when their boilers are lit prior to departure. The Clean Air Act (UK) provides special measures for such occurrences. Marine fuel tends to have a higher sulphur content and coal fired vessels were the worst offenders.

Figure 7.4 Ships in or just leaving port can be a large local source of pollution (Liverpool Docks, UK).

The typical emissions for aircraft are given in table 7.5. For aircraft cruising at high altitude (e.g. supersonic airliners), release of H_2O and NO_x may be a problem in respect of the depletion of the ozone layer. However recent results seem to indicate that the number of planes flying in the stratosphere is insufficient to warrant large scale concern.

Table 7.5   Typical aircraft emissions.

Emission (kg h^{-1})	Idling	Take off	Cruising
HC	10	0.5	1
Aldehydes	1	0	1
NO$_x$	1	2	20

ALTERNATIVE FORMS OF TRANSPORT

Present day emissions from petrol and diesel engines in the USA alone are a staggering 10^{11} tonnes CO, 3×10^{10} tonnes unburned hydrocarbons and 2×10^{10} tonnes NO$_x$. Figures for the UK are given in table 7.6. A programme of control would thus seem to be of high priority. However bearing in mind the increasing demise of oil and oil-based products, it is perhaps worth considering here some alternative forms of power for personal and commercial road transport in terms of their pollutant emissions.

Table 7.6   UK pollutant emissions from road vehicles (in thousand tonnes). From *Digest of Environmental Pollution and Water Statistics* No 4 Additional Tables (1981).

		CO	HC†	NO$_x$‡	SO$_2$	Particulates	Lead
Petrol engines	1970	5714	278	235	14	21	6.5
	1975	6473	314	266	13	24	7.4
	1980	7689	373	316	11	29	7.5
Diesel engines	1970	206	33	151	30	30	0
	1975	222	36	162	38	32	0
	1980	241	39	176	30	35	0

† Expressed as hexane and excluding evaporative losses.
‡ Expressed as NO$_2$.

The *Wankel rotary engine* is in present day use by a few manufacturers. The combustion chamber has a high surface to volume ratio resulting in very high emission rates of hydrocarbons which are easily eliminated by an afterburner.

The *stratified charge engine* has low, but variable emissions; carbon monoxide concentrations are low but NO$_x$ relatively high.

The *Stirling engine* is an external combustion engine which uses heat exchangers. Although emissions are less than for the internal combustion engine it is heavier and more expensive at its present stage of development. An external combustion engine has been tested by Fiat. The combustion chamber is spherical; the fuel can be petrol, diesel or alcohol. Fuel consumption is good (17.5 km l^{-1} or 50 mpg at 112 km h^{-1} or 70 mph are the USEPA figures).

The *Rankine cycle steam engine* has the problems of typical steam operation, namely long warmup time and large weight and size.

Electric systems are under intensive development at present. Battery operated cars are, at the moment, relatively expensive with low power and limited range, needing frequent recharging. The battery weight is large (up to 50 % of the weight of the vehicle). The advantages are however considerable. The only emissions are small, being those associated with lead-acid batteries (H_2, O_2). No inflammable petrol is carried and the vehicle needs no carburettor, air cleaner, fuel pump, water pump, starter motor or dynamo.

A hybrid engine is perhaps the near perfect environmental solution. For example a petrol/electric system would operate with its electric engine in urban areas at low speeds (when pollutant emissions are critical) and use its petrol engine for high speed cruising in non-urban areas. However the load of two engines must restrict the efficiency since either the petrol or the electric engine is always present as a dead weight. In late 1983 a fleet of ten buses and ten light vans was introduced in Manchester. These vehicles use a much lighter battery, which is re-charged whilst the vehicle is still in motion by using spare energy from the conventional engine.

In the longer term, a hydrogen based economy may be the answer. Some countries (e.g. Canada) are investigating such possibilities to supplement alcohol, natural gas, propane, petrol and oil by the end of the century. At the same time encouragement is being given in many countries to the more extensive use of bicycles and/or improved mass transit systems. Improved telecommunications also may decrease the need for personal travel.

PHOTOCHEMICAL SMOG FORMATION

Although sulphurous (reducing) smogs have been known for more than six centuries, the inception of the 'soup' of airborne pollutants (both primary and secondary) known today as a photochemical smog is of recent origin. On the west coast of America the combination of hydrocarbons, oxides of nitrogen and high irradiation (first observed in Los Angeles in the 1940s) is ideal for these sunlight-catalysed reactions (figure 7.5). Although the word smog is used, its nature is very different being primarily oxidising in character. In Los Angeles it is also exacerbated by the local topography and climate.

Athough the total description of the chemical reactions is not yet completely delineated, the basic reactions producing a photochemical smog are listed below.

When large volumes of traffic are present, there will be a large emission of NO_x (NO being the primary pollutant). Atmospheric nitrogen, from the air taken into the engine for combustion, is 'fixed' at high temperatures

$$N_2 + O_2 \longrightarrow 2NO.$$

In fact this is a slow reaction which summarises two simultaneous reactions known as the Zeldovich mechanism:

$$N + O_2 \longrightarrow NO + O$$

$$O + N_2 \longrightarrow NO + N.$$

If the effluent is cooled slowly, dissociation occurs. However, in vehicles cooling is usually rapid, and the NO is fixed and emitted from the vehicle. Only a small proportion (about 5–10 %) of the NO_x emitted from cars is in the further oxidised state of NO_2. This oxidation usually occurs in the atmosphere, making NO_2 a secondary pollutant. The overall equation is

$$2NO + O_2 \longrightarrow 2NO_2.$$

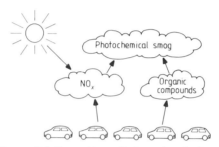

Figure 7.5 Formation of photochemical smog.

The presence of NO_2 in the atmosphere permits further reactions since the nitrogen dioxide molecule absorbs solar radiation in the 0.38–0.60 μm wavelength region. For wavelengths less than about 0.4 μm, the NO_2 molecule can be dissociated

$$NO_2 + h\nu \longrightarrow NO + O$$

where $h\nu$ represents a photon of light energy. This reaction is very rapid. In full sunlight it would deplete half the NO_2 in about two minutes. However a further reaction between molecular and atomic oxygen (together with a third body represented here by the symbol M) occurs very rapidly

$$O + O_2 + M \longrightarrow O_3 + M.$$

The species M carries away excess energy whilst remaining chemically unchanged. The ozone then reacts with the nitric oxide

$$O_3 + NO \longrightarrow NO_2 + O_2.$$

This set of reactions which is satisfied by a set of equilibrium conditions is represented diagrammatically in figure 7.6. However, calculated equilibrium concentrations fail to account for the large concentrations of ozone observed in Los Angeles. Prevalent nitrogen dioxide concentrations of 0.1 ppm should be in equilibrium with 0.03 ppm of ozone, whereas the observed concentrations were more than an order of magnitude greater. This NO_x-O_3 cycle was thus not the complete story. When carbon monoxide and water vapour are also present the NO_2-O_3 cycle may be modified:

$$
\begin{array}{rcl}
O_3 + h\nu & \longrightarrow & O_2 + O \\
O + H_2O & \longrightarrow & 2OH \\
OH + CO & \longrightarrow & CO_2 + H \\
H + O_2 + M & \longrightarrow & HO_2 + M \\
HO_2 + NO & \longrightarrow & OH + NO_2.
\end{array}
$$

This CO–OH cycle is relatively slow (Seinfeld 1980) but begins to demonstrate the presence of OH and HO_2 in the atmosphere, which react readily with other compounds. For example, the hydroxyl radical reacts directly with NO and NO_2:

$$
\begin{array}{rcl}
OH + NO & \longrightarrow & HNO_2 \text{ (nitrous acid)} \\
OH + NO_2 & \longrightarrow & HNO_3 \text{ (nitric acid).}
\end{array}
$$

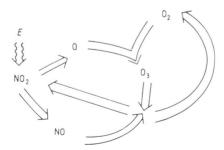

Figure 7.6 Reactions in photochemical smog.

The realisation that hydrocarbons (emitted for the reasons described in the earlier part of this chapter) also participate has helped to account more successfully for the known constituents of a photochemical smog. Several slower reactions can be identified (below). Some of the hydrocarbons and NO_2 are later removed by side reactions—reactions which create aerosols of high nitrate and organic content, 95 % of which are less than $0.5\,\mu m$ in size and thus are good light scatterers. This, together with absorption by NO_2, accounts for the hazy visibility in such conditions.

Some important reactions are outlined below. The reactivity of the hydrocarbon R depends upon its chemical structure. For example hydrocarbons with a double bond (e.g. ethene $CH_2{=}CH_2$) are more reactive.

Oxygenated hydrocarbons such as aldehydes may be formed e.g.

$$\text{formaldehyde} \qquad \underset{\displaystyle H-\overset{\displaystyle \overset{O}{\|}}{C}-H}{}$$

$$\text{acrolein} \qquad CH_2{=}CH-\overset{\overset{\displaystyle O}{\|}}{C}-H.$$

Aldehydes may be photolysed forming free radicals† e.g.

$$R-\overset{\overset{\displaystyle O}{\|}}{C}-H + h\nu \longrightarrow R. \quad + \quad H-\overset{\overset{\displaystyle O}{\|}}{C}.$$

alkyl   formyl
radical   radical
e.g. CH_3
or C_2H_5

The alkyl radical is easily oxidised by atmospheric oxygen

$$R. + O_2 \longrightarrow ROO.$$

The hydrocarbons also compete for the atomic oxygen and ozone:

$$O + \text{hydrocarbons} \longrightarrow R._1 + R-C{=}O$$

$$O_3 + \text{hydrocarbons} \longrightarrow R-C\underset{\diagdown O.}{\overset{\diagup O}{}}$$

An important reaction here is when acyl radicals $R'-C.{=}O$ (where R' is an alkyl radical) are also present. These may be formed as a result of the reaction

$$R-\overset{\overset{\displaystyle O}{\|}}{C}-H + OH \longrightarrow R-C.\overset{\diagup O}{} + H_2O$$

and are important in the conversion of NO to NO_2 e.g.

$$R'-C.\overset{\diagup O}{} + O_2 \longrightarrow R'-C\underset{\diagdown O-O.}{\overset{\diagup O}{}}$$

$$R'-C\underset{\diagdown O-O.}{\overset{\diagup O}{}} + NO \longrightarrow R'-C\underset{\diagdown O.}{\overset{\diagup O}{}} + NO_2$$

†The unused bonding electrons in the radical are shown as dots, e.g. R..

A second mechanism for this conversion is given by

$$ROO. + NO \longrightarrow NO_2 + RO.$$

When the nitrogen dioxide concentration is sufficiently high a competing reaction begins

The product of this reaction is the group of chemicals known as PAN (peroxyacylnitrate). The individual species depends upon the alkyl radical R'. Two of the more important members of this family are the simple peroxyacetylnitrate and the probable carcinogen peroxybenzoylnitrate (PBzN).

The combination of PAN, acrolein, ozone and other species is an oxidant, causing for example plant damage, material damage, discolouration of the atmosphere, eye irritation (see table 7.7) and is possibly toxic. A measure of the 'total oxidant' content of the atmosphere can be gained by bubbling a sample through potassium iodide, such that the I^- ion is oxidised.

Table 7.7 Approximate eye irritation thresholds for a five minute exposure.

Pollutant	Concentration (ppm)
Formaldehyde	1 ppm
PAN	0.7 ppm
Acrolein	0.5 ppm
PBzN	0.005 ppm

The formation of PAN and ozone is insignificant if hydrocarbon emissions are low and thus steps to reduce photochemical smog formation can be assisted by direct control of hydrocarbon emissions from vehicles (as described above).

Table 7.8 shows the emissions for Los Angeles County where 80 % of the contaminants originate from motor vehicles. This is also reflected in figures for the diel variability of observed concentrations in urban areas (figure 7.7) where the peaks are clearly associated with the morning and evening rush hours. Total oxidant also varies over a 24 hour period with ozone and aldehydes reaching a peak in mid-afternoon by which time NO_x concentrations have decreased.

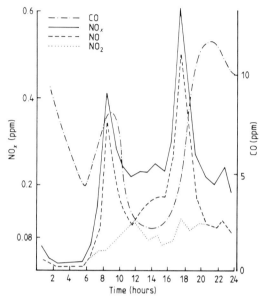

Figure 7.7 Diurnal variation of NO_x. Redrawn from Meetham *et al* (1981) by permission.

Table 7.8 Contaminants, in tonnes per day, from major sources within Los Angeles County.

Major source	Organic gases, classed by reactivity			Partic-ulates	NO_x	SO_2	CO	Total
	High	Low	Total					
Motor vehicles	1255	475	1730	45	645	30	9470	11 920
Organic solvent use	100	400	500	17	—	1	—	520
Petroleum	55	165	220	4	45	55	30	355
Aircraft	45	45	90	12	15	3	190	310
Combustion of fuels	—	9	9	15	235	40	1	300
Chemical	—	—	—	—	—	90	—	90
Other	—	3	3	16	11	3	4	35
Total (rounded)	1455	1095	2550	110	950	225	9695	13 530

SUMMARY

The role of the private car as a source of pollutants has rapidly grown. Current legislation (e.g. reduction of lead in petrol in Europe, control of hydrocarbon and other emissions in USA) is aimed at reducing the vast quantities of hydrocarbons, carbon monoxide, nitrogen oxides and lead in the atmosphere. The first three may combine under the action of sunlight (an agent presently increasing in urban areas where 'traditional' pollutants such as smoke and SO_2 have been dramatically decreased†) to form a photochemical smog including irritants and possible carcinogens. There are also indications that a synergism exists between ozone and sulphur dioxide. These are problems besetting many major cities in the world where the density of cars is high and dispersion conditions poor— problems whose incidence is likely to increase before the situation is controlled.

---

† In the UK SO_2 emissions are not controlled and thus there exists no *a priori* reason for anticipating a decrease. However a large part of the observed decrease is interpreted as being due to changing fuel usage especially from local, domestic combustion (low level sources) to regional (and often rurally sited) power stations (high level sources).

8

Air Pollution Control

Some countries set legal limits for pollutants in factories, at the point of emission or at ground level, and others have only broad guidelines. Clean Air legislation in many countries has done much to improve visibility, reduce smoky emissions etc. However, although the townscape, in terms of air quality perception, is vastly improved, there may still remain large quantities of invisible pollutants (e.g. SO_2, fine particulates). The degree of injury to health caused by these pollutants has, in many cases, not yet been fully evaluated. New concerns about their health effects, especially to man, means that increasing sophistication and care of operation in the field of air pollution control technology is demanded.

Control of pollutant emissions may be performed either within the working environment—such that workers do not become contaminated; or within the waste gas stream (e.g. before the waste gases enter the stack); or both. In any case an air pollution problem is replaced by a solid (or liquid sludge) disposal problem. Wastes emitted as part of the process and vented to the atmosphere via (usually) a tall stack may be moderated by the installation of control equipment (as described later in this chapter). These *controlled* emissions should be contrasted with *fugitive* emissions—which are less easy to eliminate, occurring as they do infrequently or at unpredicted locations, perhaps as the result of a minor accident or an equipment malfunction.

For the purposes of control, pollutants can be divided into two basic types: particulates and gases.

PARTICULATE REMOVAL

There is likely to be a great variety of shapes, sizes, densities, concentrations etc in the particles which are entrained into the waste gases leaving the combustion chamber. In some industries these factors will remain relatively constant and in others will be unpredictable. The problems of each industry are therefore likely to be particular to it. This makes it difficult to formulate general rules for the wise selection of control equipment. The source of particles will lead to decisions about local control (if human presence is necessitated near the source, for instance) or choice of large scale equipment which may need to be operated on either a continuous or batch basis. Process accidents must also be allowed for as much as possible (for example by ducting away pressure release valve emissions for further (emergency) treatment).

Control of dust in the working environment can be undertaken by the employment of either personal protection equipment, such as respiratory protective equipment (RPE) or local exhaust ventilation (LEV).

Personal protection equipment

Personal protection equipment may be in the form of masks, earplugs, gloves, full overalls or full protective suits. The particular design and degree of protection will, of course, depend upon the specific pollutant problem and may be required by law, especially if a local exhaust ventilation system is not installed. In some countries the onus on use of this equipment rests on the wearer i.e. the employee; although employers may encourage (by supplying and displaying reminder posters) strict compliance with this protection. Elsewhere (e.g. USA) the employer must enforce correct use. Mass production means that personal protection equipment (and especially RPE) is readily available. However this also implies a limited number of design types. For example, in the case of simple face masks, the multitude of facial shapes cannot be satisfied by the small range of RPE available and this often leads to neglect on the part of the employee. Particular groups of workers are likely to find the equipment uncomfortable e.g. people with beards, glasses or an unusual shape of face. Misuse (i.e. non-use) is perhaps too easy. However such disadvantages are offset to a large degree by the fact that this personal protection allows the wearer to be fully mobile and move about different parts of the plant.

Local exhaust ventilation

There are two types of ventilation that can be applied to a specific source of contaminant: dilution ventilation and LEV. Dilution ventilation is of limited applicability. In this method uncontaminated air is introduced in large volumes so that the pollutant is diluted to concentration values that are acceptable. (This method is only used for dusts that may be regarded as fire or explosion hazards at high concentrations since the procedure results in dust settling out causing localised 'housekeeping' problems.) The air may be introduced either by natural or mechanical means. Often the use of dilution ventilation is augmented by the use of RPE as well.

Local exhaust ventilation is a method of capturing the dust at or near the source and then containing it and removing it from the working environment. The parts of the system are: entrance; duct; contaminant remover;

air mover. (The order is analogous in many ways to a sampling train for pollutant measurement—see Chapter 4.)

The entrance is usually in the form of a hood (figure 8.1). The design of the hood is critical. Each individual case needs separate consideration. The collection can be reinforced by the use of partial or total enclosures e.g. booths. Particles are then drawn from the source by the air flow created by the air mover and collected by the hood. Hence the operator must never place himself between the source and the offtake. Although there is a multitude of designs for hoods, they fall broadly into two categories: receptor and captor hoods.

Figure 8.1 Hood for removal of waste gases (here for removal of gases from an AA spectrophotometer).

A *receptor hood* receives a dust cloud which is driven into it by external forces e.g. the kinetic energy originating from an explosion. The hood must be emptied as quickly as it is filled to maintain efficiency. For a *captor hood* the dust velocity is zero and thus the LEV system itself must increase the velocity of the dust. The simplest form for a captor hood is an open duct. A low pressure is maintained inside the hood. Particulate pollution, removed from the working environment, or indeed pollutants created elsewhere (e.g. in a furnace), will eventually be transported away via a gas stream.

Before final emission, often via a tall stack, it may be expedient to pass the gas stream through some form of air pollution control device to limit the atmospheric pollution (and possibly even to recover the dust for other uses). At this stage there are *four* basic types of control equipment: mechanical collectors, fabric filters, wet scrubbers and electrostatic precipitators.

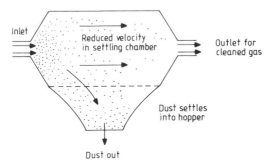

Figure 8.2 Gravity settling chamber.

Mechanical (or inertial) collectors
Any inertial collector relies on the effect of gravity acting differentially between particles and carrier gas and/or the inertial effect in which the drag on the particle is greater than on the carrier gas. The simplest form of mechanical collector is the settling chamber (figure 8.2). The dust-laden gas is introduced into a large chamber via a narrow nozzle. Continuity demands that the change in cross-sectional area of the gas flow this induces causes a decrease in the velocity of the gas stream. Whilst the gas travels slowly through the chamber the effect of gravity on the larger, denser particles is to cause them to settle downwards. This settling velocity can be calculated by using Stokes's law (see Chapter 1) typically in the range $0.3-3.0 \ \mathrm{m \, s^{-1}}$ and thus the time taken for a settlement over the vertical distance, defined by the depth of the settling chamber, can be calculated. It is therefore necessary to ensure that the residence time of the dust in the settling chamber is sufficiently long for this settlement to occur. Dust collected in the hoppers at the base of the chamber can thus be removed and disposed of as a solid waste. The basic characteristics of the gravity settling chamber are simplicity of construction and maintenance (and hence low costs). Energy usage is also low. Although the reliability is excellent, collection efficiency is low and this decreases at high loading. For fine and moderately fine dusts gravity settlement cannot be recommended. Dust settlement can be encouraged by a modification to the basic design. Figure 8.3 shows one such modification—a recirculating baffle collector. This has several advantages over the gravity settlement chamber. The alteration to the direction of the gas flow through the chamber has a portion where the gas flow is

vertically downwards. As the trajectory and resultant velocity of the particles depend upon the strength and direction of the flow as well as the settling velocity, there is more rapid settlement in this part of the chamber. Additionally this sudden change of direction utilises the inertia of particulates. They will impinge upon the walls and lose some of their velocity—again an encouragement to settlement.

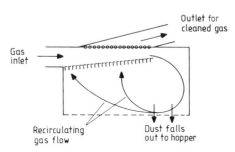

Figure 8.3 Baffle collector.

A very common form of inertial collection is to be found in the *cyclone* collector (also known as a centrifugal collector). The basic concepts are exemplified in figure 8.4. The dust laden air enters at the top and spirals slowly down. The ascending high velocity core creates a centrifugal force acting on the dust particles which makes them move towards the outer walls. This force (F) is given by

$$F = MV^2/R \tag{8.1}$$

in which M is the mass of the particular particle, V is its velocity and R the cyclone radius. These are the three main variables which determine the efficiency of operation of a cyclone. A close inspection of equation (8.1) will show for which particles cyclone collection is best suited. For example either increasing the particle mass or decreasing the cyclone radius will increase the collection efficiency. Similarly performance can be improved by an increase in gas velocity up to a value at which scouring becomes excessive. Although there are several types of cyclone they fall into two groups dependent upon the mode of introduction of the gas stream which can be either by means of a tangential inlet or an axial inlet. The former are usually fairly large (1–5 metres in diameter), whereas axial inlet cyclones are usually much smaller (≤ 2 metres in diameter) and are consequently often used in a bank of parallel units. The cyclone principle is, in practice, usually used for removal of particulates in the size range 10–100 μm. As emission standards are tightened it is likely that the cyclone will be used less as a result of its low collection efficiency for very small particles (see figure 8.5). However it is relatively simple in construction and reliable in operation.

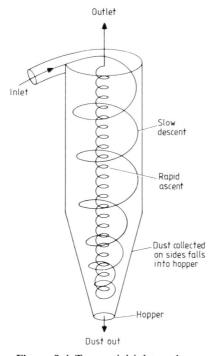

Figure 8.4 Tangential inlet cyclone.

In addition to these types of mechanical collector there is a range of further devices in less common usage. Some examples are cinder traps (effectively a gravity settlement plus multiple baffles), louvre separators, dynamic precipitators and wet cyclones (in which capture is increased by spraying water into the inlet). This latter method may also be used to keep down the temperature of a hot effluent.

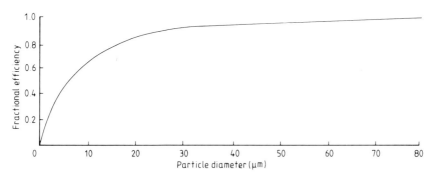

Figure 8.5 Typical cyclone fractional efficiency curve.

Filters

Fabric filters are made from suitable synthetic or mineral fabric or felt. They are housed in a baghouse and collect dust by several methods (see figure 8.6). There are three main ways of collection: interception, electrostatic attraction and inertial impaction. Particulates carried into the filter along with the gas stream will have trajectories following the gas streamlines until the gas flow is deviated. The larger density of the dust is reflected in its greater inertia and resistance to directional changes. The dust particles are thus not deviated around the filter element, as is the gas stream, and may collide with the filter element (inertial impaction). If the particle is being carried along towards the element, on or almost on the axis of the filter element, then collection is by direct impaction (or interception). Electrostatic attraction also may remove smaller particles from the streamline along which they are being carried by the gas flow.

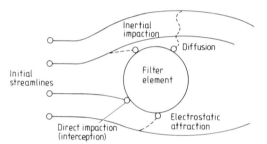

Figure 8.6 Collection mechanisms for a fabric filter.

There are, in addition, some other minor removal mechanisms. For larger particles, the mesh of filter elements can act as a physical sieve. Some settlement may be induced by gravity and there may be minor lateral diffusion effects (more important for smaller particles) or thermal precipitation due to any existing temperature gradient between the hot gas stream and the (possibly cooler) filter element.

Impaction does not of course necessarily mean collection. Due consideration is needed for the efficiency of adhesion for particle impaction for each of the methods outlined above. Even if a particle adheres it may be dislodged at a later time by dust particles arriving. Consequently as a fabric filter is being used, the deposit on its elements increases and eventually the filter must be cleaned (either on or off line) or replaced.

On a commercial basis the filters are housed in a baghouse. Although highly efficient at most gas flows there are associated high maintenance costs due to bag replacement. Some *in situ* cleaning can however be successful. Mechanical shaking, reverse air pulses, continuous reverse air jets and collapse/pulsation methods can all be used.

The filter elements themselves may be made from cotton, glass fibre or nylon as well as the more traditional fabrics. Although the interstices between elements may be as large as 100 μm, they are able to remove particles of sizes down to about 0.5 μm.

Other types of filtration are currently under development. These include the use of fluidised beds, rigid porous materials such as porous ceramics and plastics, paper filters and throwaway filters. One development now in use in many places is the high efficiency particulate filter (HEPA). They too remove dusts by inertia, diffusion and interception and are best used for small particles often as a pretreatment before passing the gas flow on to an ordinary fabric filter. Their construction is of a fibrous mat of sub-micron sized fibres which have little deviating effect on the streamlines. Hence interception is the main removal mechanism. To enhance this the filter material (often paper) is pleated to increase the effective cross sectional area presented to the flow. By continuity, this can be seen to result in a decreased velocity across the filter and hence more effective capture as a result of the increased inertial impaction. They are highly efficient especially for the toxically most dangerous dusts—those below about 2 or 3 μm in diameter, sizes at which bag filters are notoriously inefficient. Some HEPA filters can be self-cleaning; others form a subset of 'absolute filters', so called because their performance complies with a test using a specified and well determined aerosol.

Wet scrubbers

A wet scrubber is a device which uses a liquid to capture and hence remove particulates from a moving stream of gas. This can be accomplished in two basic ways: (i) in which the dust is forced to impinge on a wetted surface or (ii) in which a fine spray of liquid droplets captures the suspended dust particles and brings them (by gravity) to a collection area. In this maximum efficiency is achieved when the relative velocity between the dust and the liquid drops is as high as possible. There are several designs of wet scrubbers; some of which will be discussed briefly below.

A gravity settling chamber (see figure 8.2) to which is added a fine spray (usually of water) is perhaps the simplest type of wet scrubber. The slowing of the gas as it enters the chamber again assists in the settlement. Most of the cleaning is done by collision and the inclusion of a demist section will ensure that the maximum settlement of both dust and water is attained. The efficiency of operation of such a wet scrubber can be as high as 99 %. There is however a possibility of corrosion and there is likely to be a white water laden plume emitted at the stack top. A development of this basic idea is shown in figure 8.7. This is the wetted impingement baffle scrubber in which the inclusion of a series of baffles enforces several changes of direction of the air flow and hence better settlement. The venturi effect can also be utilised in scrubbers and such devices can be very compact.

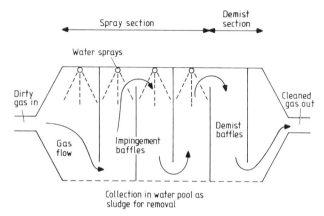

Figure 8.7 Wet impingement baffle scrubber.

Electrostatic precipitation

Electrostatic precipitators are in common use especially when fine dust would otherwise be a problem. They were invented at the University of California by Frederick Cottrell in 1910 and developed originally for the recovery of process material for reuse. They are able to handle high temperature gases with ease which makes them ideally suited for boilers and furnaces. There is only a small pressure drop across the device so fan costs are minimised. They have a high collection efficiency if operated on selected aerosols especially for size ranges below about $10\,\mu$m. They are, however, bulky, have high capital and running costs and are not appropriate for combustible particles such as grain and wood dusts.

Their operating principle is outlined in figure 8.8. There are six steps in the Cottrell process, the first four being continuous, the latter two intermittent at intervals of between a minute and several hours. A DC potential of about 40 000 to 50 000 volts below ground potential exists between the middle wire and the grounded plate (the collecting electrode or 'collectrode'). This causes the intervening gas to become ionised such that an electric 'wind' is created of negatively charged gas ions migrating from the wire to the wall. Any dust in this air then becomes negatively charged and precipitates on to the positively charged collectors. Step (4) shows the dust layer building up. As this layer thickens the negative charge is slowly lost and eventually the incoming dust is no longer retained by the (now insufficiently strong) electrostatic charge. The dust resistance depends not only on the layer thickness but also on the type of dust. The measurement of the 'dust resistivity' is often used as an indicator of the potential use of the electrostatic precipitator in any specific case. When the dust layer is 1.5–6 mm in thickness collectrode rapping is undertaken to dislodge the dust layer which will then fall in relatively large agglomerates into the hoppers.

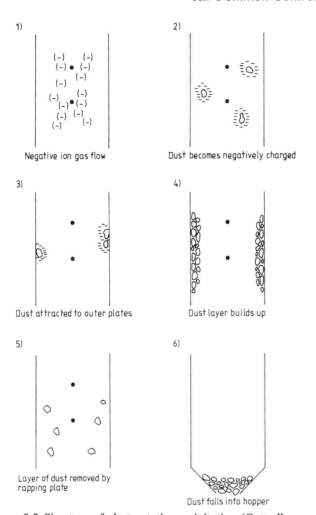

Figure 8.8 Six steps of electrostatic precipitation (Cottrell process).

GASEOUS TREATMENT

The presence of waste (and undesirable) gases in the effluent stream may cause a bigger problem in determining the treatment process required. Gases cannot be separated by any of the processes described above which rely on inertial properties. Separation must rely on the chemical and/or physical characteristics of the gas. In general gaseous treatment can be subdivided into four basic methods: absorption, adsorption, condensation and conversion.

Absorption

Absorption is the gaseous analogy of the wet scrubber for particulates and often utilises treatment plant of very similar design. The waste gas stream is passed through a liquid medium in which it has a high solubility. Removal will be accomplished either by direct solution or by solution following a chemical reaction. For example, scrubbing a waste gas containing HCl with water creates a solution of hydrochloric acid. Scrubbing the same stream with NaOH (in weak solution) results in a chemical reaction which neutralises the acidic gas at the same time as removing it by absorption. Many gases can be treated in this way e.g. SO_2, H_2S, HCl, Cl_2, NH_3, NO_x and HC. The solvent is often water although it must be chosen so that the solubility is high, its vapour pressure is low (to reduce losses), it is inexpensive, non-toxic, non-flammable and chemically stable with a low freezing point.

Absorption equipment may be a packed column, plate column, spray tower or many other variants on these basic designs. Although the contaminant is removed, large volumes of water vapour are often added, and when emitted from a chimney this causes a visible plume (figure 3.12), which is cooler than the unscrubbed effluent and, being less buoyant, is thus likely to reach the ground earlier (and hence in a less dilute state). However the advantage of the cleaner effluent is usually beneficial (except perhaps in unstable, looping plume conditions).

Wet scrubbing techniques are often used for SO_2 removal from, especially, power plant effluents. Such flue gas desulphurisation (FGD) has the disadvantages (outlined above) of producing a cool, saturated plume and reheating is often needed. These problems have led to the recent investigation into the feasibility of 'dry' FGD. In this developing technology the flue gas is brought into contact with a lime slurry such that a dry solid is produced. The remaining gaseous effluent is neither wet nor cool and it becomes feasible to have 'stackless' power stations in which the FGD plant is contained within the cooling (e.g. Esche 1983).

Adsorption

Adsorption describes the process by which gaseous molecules are attracted to and retained on the surface or in the interstices of solid particles. Again this may be a simple surface phenomenon in which retention is by molecular forces and/or it may include a chemical reaction in which case the true terminology is *chemisorption*. The solid used must have a high surface to volume ratio, be capable of being packed and must not fracture easily. If at all possible it must also be capable of regeneration.

Initially the process is nearly 100 % efficient since there is a large surface area available for the adsorption. As the surface is covered or the interstices filled, the efficiency falls. This is observed to occur relatively suddenly. Figure 8.9 shows how the efficiency drops once the *breakpoint* is reached.

Industrial units must be regenerated before this time since later the efficiency of removal drops rapidly to zero and uncleaned gas passes through the cleaning device to be vented to the atmosphere. Activated carbon is very useful in this context, especially for odourous light hydrocarbons, with a surface of about 1.4×10^6 m^2 kg^{-1}. Alumina is useful for the chemisorption of sulphur dioxide and silica gel is well known for its ability to remove water vapour.

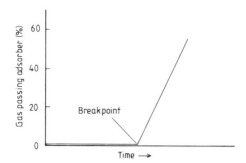

Figure 8.9 Efficiency of adsorption as a function of time.

Condensation

In some situations a condensation may be desirable, often ahead of other control equipment, for example, when the gas to be condensed is potentially corrosive. Condensation is also used when the byproduct has a potential economic value for reuse or sale. It is also an invaluable means of reducing the total volume of gas to be disposed of and hence in the saving of pipework, fans, pumps etc. Condensation is undertaken by one of two methods: reduction of the temperature or increase of the pressure. The latter method is seldom used because of the high construction and operating costs of high pressure vessels. For temperature reduction two types of condenser are in common use: contact and surface. Figure 8.10 shows a contact condenser in which water is sprayed into a chamber into contact with the waste gas stream rising through the vessel. The mixing of the coolant and gases cools the gas stream such that the temperature of the vapour to be removed falls below its boiling point and condensation occurs. In a surface condenser the heat exchange occurs across a metal surface (i.e. the coolant and the gas do not come into contact). The heat exchanger can be of either a shell or a tube type. Condensers are much used in petroleum refining, ammonia and chlorine manufacture, dry cleaning and in the Kraft paper process. Operating temperatures are usually such that the vapour temperature is reduced from about 373 K (100 °C) to about 293–298 K (20–25 °C). Thus gases for which this type of removal would be applicable are those which are in the gaseous phase at the higher temperature and the liquid phase at the cooler temperature.

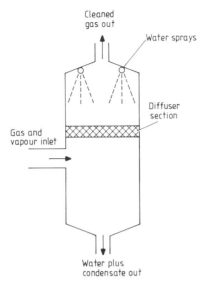

Figure 8.10 A contact condenser.

Conversion

In some instances it is perhaps useful to convert the pollutant gas to a non-pollutant (or in some cases to a less harmful pollutant). A common use is in the oxidation of organics to carbon dioxide and water. In this process, known as *afterburning*, combustion is initiated in a second combustion chamber in either (i) a direct system or (ii) a catalytic system which operates at a considerably lower temperature. Direct flame afterburners are most common for vapours and gases operating at a temperature of 925–1100 K, with a retention time of 0.3–0.5 s at the higher temperature. If properly operated they can be nearly 100 % efficient. Often heat recovery is possible and economic. If a lower temperature is required the use of a catalyst may be considered. This is useful for the control of solvents and organic vapours. Catalytic afterburners have a great potential for the control of exhaust emissions in cars (see Chapter 7). Their operating temperature is in the region of 600–800 K. Installation costs are higher (the catalyst is usually expensive and needs regular replacement due to gradual poisoning). Optimum operation is determined, as before, by time, turbulence and temperature.

Conversion is sometimes used for the treatment of a gas stream containing hydrogen sulphide. Oxidation of this results in the creation of sulphur dioxide. Although this is also regarded as a pollutant, it is less toxic than the original hydrogen sulphide (the odour threshold for H_2S being three orders of magnitude less than for SO_2). The sulphur dioxide can subsequently be removed by, for example, scrubbing.

LEGAL REQUIREMENTS

Legislative control is designed to restrict emissions and minimise ground level concentrations, whilst permitting industry to function at an economic cost, if at all possible. To attain this goal, countries have adopted different control strategies which broadly fall into two groups: emission standards or 'best practicable means'. Typifying these methods we will discuss the USA and the UK. The former approach produces a figure for the emission of a number of pollutants and is usually linked to an ambient concentration standard. In the latter approach, emissions are restricted such that ground level concentrations do not rise appreciably when new industry is introduced into an area. In practice, controls are only introduced as cost and available technology will permit. If there are no 'practicable means' by which these limits may be reached, standards may be ignored or the law becomes relaxed or deferred until such technology becomes available. Thus even an 'emissions standards' approach may be strongly moderated by the best available practicable means.

In addition, the TLVs (see Chapter 5) are in use in the UK, the USA and many other countries.

Table 8.1 USA Federal standards for ambient air quality.

Pollutant	Annual mean †	24 h max	8 h max	3 h max	1 h max
Primary standards (μg m^{-3} (ppm), values not to be exceeded more than once per year)					
SO_2	80(0.03)	365(0.14)			
CO			10 000(9)		40 000(35)
NO_x(as NO_2)	100(0.05)				
Ozone					160(0.08)
HC(as CH_4)				6–9 AM 160(0.24)	
Particulates	75	260			

Secondary standards (μg m^{-3} (ppm), values not to be exceeded more than once per year)
As above except for SO_2 and particulates:

SO_2				1300(0.50)	
Particulates	60 (guideline)	150			

† Annual mean is arithmetic for SO_2, NO_x and geometric for particulates.

United States legislation

Emission standards are usually defined by each state, although guidelines are given by the USEPA formed in 1970. At the same time a set of federal standards for ambient concentrations has been in force (table 8.1) since the Clean Air Act Amendments of 1970. These are expressed in terms of a primary standard (for general health protection) and a secondary standard (for protecting against other known or anticipated adverse effects). The dates for compliance with these were 1975 (primary standards) and 1977 (secondary standards). In any cleaner areas where concentrations are lower than these standards the concept of 'prevention of significant deterioration' (PSD) is used. The values given in table 8.2 give the incremental limits allowable before it is considered that a significant deterioration in the atmospheric environment has occurred. Proposals for new industry in such areas must include the use of 'best available control technology' (BACT). When new industry is proposed (especially in Class 1 areas) an environmental impact assessment (EIA) may be required. This is undertaken by considering the worst case that would occur. To do this meteorological data are used to calculate resultant ground level concentrations and hence evaluate the impact of the new source.

Table 8.2 Prevention of significant deterioration (PSD) SO_2 air quality standards (USA). Values not to be exceeded more than once per year.

USEPA classified areas	Averages (μg m^{-3})		
	Annual	24 h	3 h
Class 3: Industrial	80	365	1300
Class 2: Most of USA	15	100	700
Class 1: Pristine (e.g. national parks, wilderness, Indian lands)	2	5	25

Restrictions for vehicle emissions were brought into force by state and federal legislation (e.g. Motor Vehicle Air Pollution Control Act (1965), the effects of which have been discussed in Chapter 7). The 1977 Clean Air Amendments required the following emission standards to be met:

HC	120 ppm
CO	6400 ppm
NO$_x$	550 ppm.

These have been progressively reduced. By 1980 the hydrocarbon limit was 0.41 g mile^{-1} (0.25 g km^{-1}) and CO 7 g mile^{-1} (4.35 g km^{-1}).

United Kingdom legislation

The concept of 'best practicable means' arose with the formation of the Alkali and Clean Air Inspectorate (ACAI) in 1863. Today the ACAI† is responsible for regulating emissions from 61 'registered' industrial processes. Together with the Factory Inspectorate, responsible for pollutant levels

†Since 1983 renamed the Industrial Air Pollution Inspectorate.

within the (enclosed) working environment, it has since 1974 been part of the Health and Safety Executive (HSE). Other sources are controlled at local level, largely by Environmental Health Officers. The relevant pieces of parliamentary legislation are the Clean Air Acts of 1956 and 1963, the Control of Pollution Act (1974) and the Health and Safety at Work Act (1974). Smoke emissions are restricted to short periods, e.g. lighting up, soot blowing, using the Ringelmann shade (figure 4.13) of the smoke as a guide. Chimney heights are often calculated using the Chimney Height Memorandum. This is a nomogram which takes into account in a purely empirical way the sulphur content of the emission and the proximity and size of nearby buildings, but not the topography nor the meteorology—see Chapters 2 and 3. The Chimney Height Memorandum is based on Sutton's (1947) equation together with the Bosanquet *et al* (1950) assessment of plume rise to produce an 'uncorrected chimney height' to give maximum ground level concentrations of 0.16 ppm SO_2 at a 'typical' wind speed. Recently (1982) this Memorandum has been revised to cover a wider range of boiler sizes and to include some equations (in addition to the nomogram) to permit solution by calculators. A similar nomogram has recently been proposed for use in Switzerland. The local authority is also empowered to declare Smoke Control Areas in which only authorised smokeless fuels may be burned in the domestic grate. Local authority grants are usually available to permit the necessary changeover. Figures for the end of 1980 show a national (UK) average of 43 % for the number of domestic premises in smokeless zones, the highest being in London (94 %).

The degree of control equipment, necessary chimney height etc is thus determined by available technology, type of emission and necessity of operation of the industry in that location (e.g. value to the economy). As part of the best practicable means guidelines, a set of presumptive limits have been introduced. These are specified (see e.g. Nonhebel 1981) for each industry in the reports of the ACAI/Industrial Air Pollution Inspectorate.

European and other legislation

Although each country in the EEC has its own legislation, there are at present EEC standards being gradually imposed throughout the Community. The Directive 80/779/EEC gives values for smoke and SO_2 as an annual value (median), winter median and annual 98 percentile. For each value the smoke limit is set together with a sulphur dioxide value which is dependent upon the ambient smoke concentration (table 8.3).

Table 8.4 gives the standards in use in 1978 (Jarrault 1978) in both individual European countries and elsewhere. It should be noted (Rubin 1981) that although total suspended particulate values differ the different methods of assessment used may differ by a factor of up to 4. Of the European countries only West Germany imposes emission limits (as opposed to ambient concentration values). The USA also imposes such limits.

Table 8.3 EEC Directive limit values (corrected to be compatible with UK monitoring methods).

	Smoke (S_1) Limit value (μg m^{-3})	Sulphur dioxide	
		Smoke range	Limit value (μg m^{-3})
Year (median of daily	68	$S_1 < 34$	120
values)		$S_1 \geqslant 34$	80
Winter (October–March)	111	$S_1 < 51$	180
(median of daily values)		$S_1 \geqslant 51$	130
Annual (98 percentile of	213	$S_1 < 128$	350
daily values)		$S_1 \geqslant 128$	250

SUMMARY

Control of pollutant emissions can be divided into technological and legislative. In many cases available technology may not be implemented because of the high capital investment needed which may make the process unprofitable. Technology has been described in terms of the removal of particulate and gaseous pollutants. Some aspects of recent legislation in the developed world have been summarised to indicate some of the possible methods of controlling air pollution.

Table 8.4 Ambient air quality standards for SO_2 and total suspended particulates (TSP) in 1978 (μg m^{-3}). Adapted from Rubin (1981).

Country	SO_2 Annual	24 hr av.	Other	TSP Annual	24 hr av.	Other
Canada (secondary standard)	60 30	300 150	900 (1 hr av.) 450 (1 hr av.)	70 60	120	
Finland	180	250	720 (30 min av.)	180	250	720 (30 min av.)
France		250			350 (special zones)	
Federal Republic of Germany			140 (long term av.) 400 (short term av.)			200 (long term av.) 400 (short term av.)
Italy		260	650 (30 min av.)		300	750 (2 hr av.)
Japan		100	260 (60 min av.)		100	200 (60 min av.)
Norway		200	60 (6 mth av.) 400 (1 hr av.)		120	40 (6 mth av.)
People's Republic of China	150			150		
Spain	150	400	256 (1 mth av.) 700 (2 hr av.)	130	300	200 (1 mth av.)
Sweden		300 (max) 200 (long term av.)	60 (6 mth av.) 750 (1 hr av.)		120	40 (6 mth av.)

Appendix

THE PERIODIC TABLE

The periodic table (figure A.1) is the skeleton for describing the similarities and differences between chemical elements. A mass of any element is composed of a large number of atoms (for example 12 kilograms of carbon contain 6.023×10^{26} atoms). Each atom of the same isotope (see below) of an element can be considered to be identical. The Bohr model of the atom (sufficient for undestanding of air pollution chemistry) is shown in figure A.2. The central part or *nucleus* contains positively charged *protons* together with uncharged *neutrons;* each of unit mass (actually 1.673×10^{-27} kg). All atoms of an element contain the same number of nuclear protons but the number of neutrons may be different. Atoms containing the same number of protons but different numbers of neutrons are *isotopes* of the element. In fact it is the number of protons, known as the *atomic number* which uniquely identifies the element; for example hydrogen (one proton), oxygen (eight protons). The total of the protons and the neutrons is the *mass number.* The mass number may be indicated by the following symbols: e.g. carbon with mass number 12 units is written ^{12}C; with a mass number of 14 units ^{14}C (read as carbon−14). Another commonly used scale is that of *atomic weights* or relative atomic mass which is the ratio between the weight of an atom and one twelfth of the weight of an atom of the isotope ^{12}C. (This replaced the scale where hydrogen has a mass of exactly 1. On the carbon scale the atomic mass of hydrogen is 1.01.)

In an electrically neutral atom, the number of protons is equalled by the number of negatively charged orbiting *electrons,* arranged in concentric 'shells' each of which has a maximum electron capacity. The filling of a shell corresponds to moving from left to right in the periodic table. For the lower elements in the periodic table, when the outer shell is full, extra electrons are located in the next outermost shell and we have moved on to the next row in the table. For example, on row 3, sodium (Na) has one outer electron, magnesium (Mg) two and so on, to a full shell (8) of argon (Ar);

adding a further electron necessitates the use of an extra shell. The outer two shells now contain 8 and 1 electrons respectively—the element is potassium (K). Hence all elements in any vertical column have the same electron configuration in their *outer* shell. Consecutive shells are composed of a number of subshells which are identified by a series of letters. These are given in table A.1, together with the maximum capacity of each shell and subshell.

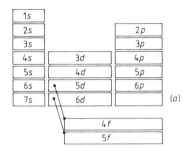

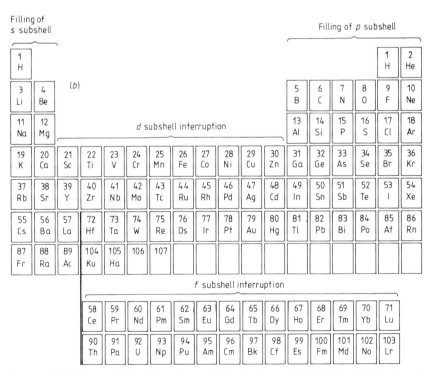

Figure A.1 The periodic table and its relationship to subshell filling. (*a*) Basic subshell structure of the periodic table. (*b*) Subshell structure with elements filled in, yielding the full periodic table.

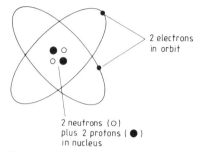

Figure A.2 The Bohr model of the atom. The example shown is helium.

Table A.1 Electron capacity of subshells.

Shell	Subshells				Total
	s	p	d	f	
1	2				2
2	2	6			8
3	2	6	10		18
4	2	6	10	14	32
5	2	6	10	14	32
6	2	6	10	14	32
7	2	6	10	14	32

The number of electrons in the outer shell is related to the *valency* or ability to give or receive electrons. Elements in the columns form closely-knit 'families' or *groups* in which the chemical properties are similar, e.g. the halides (fluorine, chlorine, iodine and astatine). Horizontal rows of the periodic table are called periods. Macroscale properties change similarly. Moving from left to right it is observed that:

(*a*) metals in group I give way gradually to non-metals in groups VI, VII and inert gases in group VIII;

(*b*) boiling points and densities increase to a maximum and then decrease;

(*c*) metals lose electrons easily (low electronegativity), non-metals gain electrons easily (high electronegativity).

Moving down a group it is found that:

(*a*) boiling point decreases (metals)/increases (non-metals);

(*b*) density increases;

(*c*) atom size increases.

Combinations of atoms of similar or different species are *molecules* and the *molecular mass* is the sum of the individual atomic masses. The molecular mass expressed in grams is known as 1 *gram-molecule* or 1 *mole* and always contains 6.023×10^{23} (Avogadro's number) molecules. The mole is a handy measure, for instance, in evaluating weights of chemical reactants, or for calculating solution strengths.

CHEMICAL BONDING

In a chemical reaction outer (valency) electrons may be exchanged to form *ions* (electrically charged atoms) or *molecules* (combinations of two or more atomic species). There are several different types of bonding of importance in atmospheric chemistry; *ionic* or *electrovalent* bonding occurs when electrons are donated or received and ions are formed. For example a reaction between sodium and chlorine results in the chlorine atom gaining an outer electron from the sodium (which now has no outer electrons and a single positive charge). The sodium *cation* is thus held electrically to the negatively charged chlorine *anion*. (Atoms with fewer than four outer electrons tend to lose them, atoms with more than four to gain them.) The resultant sodium chloride (the 'ide' indicates that no other chemical species are involved) can be written symbolically as

$$Na^+ \; Cl^-$$

Indicating the whereabouts of outer electrons by dots we can write

$$Na\,\overset{.}{} + :\overset{..}{\underset{..}{C}l}. \quad \longrightarrow \quad Na \; :\overset{..}{\underset{..}{C}l}:$$

This demonstrates the tendency for atoms to gain, by some means, a complete outer shell of eight electrons.

This shell filling is utilised in *covalent* bonding. However, electrons are no longer donated to create ions, but donated and *shared*. For instance, two hydrogen atoms form a hydrogen molecule:

$$H\,\overset{.}{} + H\overset{.}{} \quad \longrightarrow \quad H:H$$

Effectively the *shared pair* of electrons completes the outer shell (the s or innermost subshell is complete with two electrons) for *both* atoms. A shared pair or covalent bond can be indicated by a single line

$$H-H$$

Molecules can also be formed by the sharing of two pairs of electrons. Oxygen has six outer electrons and thus requires two to complete the shell. If these are supplied by another oxygen atom, then four electrons (two pairs) are shared. Symbolically we have for molecular oxygen, O_2, (the subscript is a simple summation without reference to the electronic configuration):

$$:\overset{..}{O} + \overset{..}{\underset{..}{O}}: \quad \longrightarrow \quad \overset{..}{O}::\overset{..}{\underset{..}{O}} \quad \text{or} \quad O=O$$

This may be expressed by assigning a valency or oxidation number of two to oxygen.

Some molecules cannot be attributed a unique electronic configuration —they exhibit *resonance*. For example sulphur dioxide, SO_2, could be given by

$$.\ddot{S}.\qquad\qquad .\ddot{S}.$$
$$.\ddot{O}:\quad .\ddot{O}. \qquad or \qquad :\ddot{O}. \quad .\ddot{O}:$$

Resonance does not imply that two forms exist for SO_2, merely that the molecule is a hybrid of both structures concurrently and no single structure can be identified. Symbolically

$$O \overset{-}{\underset{}{\diagup}} \overset{S}{\diagdown} O \quad \longleftrightarrow \quad O \diagup \overset{S}{\diagdown} O$$

Hydrogen bonding can be of some importance. It is not a physical bonding but an electrostatic attraction between neutral molecules in which there is a slight charge separation. For instance, water is a dipole molecule:

$$\underset{H \quad +}{\overset{-}{\diagup O - H}}$$

in which the oxygen is marginally negative and the hydrogen positive. Attraction between the hydrogen and the oxygen of a neighbouring molecule can hold together several, otherwise neutral, molecules. In fact it is hydrogen bonding of n molecules of H_2O (molecular weight 18) effectively into a molecule nH_2O that is responsible for many of the observed anomalous properties of water in which the water molecule behaves as if it were heavier than its simple formula H_2O suggests.

For many purposes it is unnecessary to specify the bonding or the structure of the compound. In this case the sum total of the atoms of each element in the compound is denoted by subscripts. For example, water consists of two atoms of hydrogen bonded with one atom of oxygen and can be denoted by H_2O. Similarly sulphur dioxide is SO_2. There are occasions with complex organic molecules when a simple summation disguises the structure in a way that makes prediction of properties impossible. For example, a compound containing three carbon atoms, four hydrogen atoms and one oxygen atom could be written as C_3H_4O but if it contained a bonding such as

$$-O-H$$

then it would have the properties of a hydroxide and ought to be written as C_3H_3OH.

The molecular weight of a compound is often required and is calculated by summing the individual atomic weights. So for C_3H_4O the molecular weight is $(3 \times 12) + (4 \times 1) + 16 = 56$. Long chain hydrocarbons and other organics have high molecular weights and this has a direct bearing on the physical properties of the compound.

References and Bibliography

Abdel Salam M S, Farag S A and Higazy N A 1981 Smoke concentrations in the Greater Cairo atmosphere *Atmos. Environ.* **15** 157–61

Adams R L 1981 Design and operation of a pulse jet baghouse system on a 70-ton arc furnace *J. Air Pollut. Control Assoc.* **31** 1058–60

Alexander R W 1982 The interpretation of lichen and fungal response to decreasing sulphur dioxide levels on Merseyside *Environ. Educ. Inf.* **2** 193–202

Allaby M and Lovelock J 1980 Spray cans: the threat that never was *New Scientist* **87** 221–4

—— 1980 Wood stoves: the trendy pollutant *New Scientist* **88** 420–2

Anfossi D 1982 Plume rise measurements at Turbigo *Atmos.Environ.* **16** 2565–74

Annand W J D and Hudson A M 1981 Meteorological effects on smoke and sulphur dioxide concentrations in the Manchester area *Atmos. Environ.* **15** 799–806

Annual conferences of the National Society for Clean Air

Annual reports on Alkali etc (London: HMSO)

Apling A J, Sullivan E J, Williams M L, Ball D J, Bernard R E, Derwent R G, Eggleton A E J, Hampton L and Waller R E 1977 Ozone concentrations in South-East England during the summer of 1976 *Nature* **269** 569–73

Ashenden T W and Mansfield T A 1978 Extreme pollution sensitivity of grasses when SO_2 and NO_2 are present in the atmosphere together *Nature* **273** 142–3

Atkins M H, Lowe J and Lewis D 1980 Air pollution and human health: a case study of environmental benefit estimation *Int.J.Env.Studies* **16** 29–34

Auliciens A and Burton I 1973 Trends in smoke concentrations before and after the Clean Air Act of 1956 *Atmos. Environ.* **7** 1063–79

Bach W 1972 *Atmospheric Pollution* (New York: McGraw-Hill)

Bailey D L R and Clayton P 1982 The measurement of suspended particle and total carbon concentrations in the atmosphere using standard smoke shade methods *Atmos. Environ.* **16** 2683–90

Ball D J and Bernard R E 1978 Evidence of photochemical haze in the atmosphere of Greater London *Nature* **271** 733–4

Ball D J and Hume R 1977 The relative importance of vehicular and domestic emissions of dark smoke in Greater London in the mid–1970s *Atmos. Environ.* **11** 1065–73

Barnes R A and Eggleton A E J 1977 The transport of atmospheric pollutants across the North Sea and English Channel *Atmos. Environ.* **11** 879–92

Beaver Committee on Air Pollution *Interim Report* December 1953 and *Report* November 1954 (London: HMSO)

Benarie M M 1980 *Urban Air Pollution Modelling* (London: Macmillan)

Bennett M 1980 Development of a Gaussian plume model appropriate to an urban area in *Atmospheric Pollution 1980* ed M M Benarie (Amsterdam: Elsevier) pp 63–8

Bennett M and Saab A E 1982 Modelling of the urban heat island and of its interaction with pollutant dispersal *Atmos. Environ.* **16** 1797–822

Bennett R J, Campbell W J P and Maughan R A 1976 Changes in atmospheric pollution concentration in *Mathematical Models for Environment Problems* ed C A Brebbia (London: Pentech Press)

Bernstein R D and Oke T 1980 Influence of pollution on urban climatology in *Advances in Environmental Science and Engineering* vol 3 ed J R Pfafflin and E N Ziegler (New York: Gordon and Breach) pp 171–202

(The) Boiler Operator's Handbook 1969 (NIFES)

Bosanquet C H, Carey W F and Halton E M 1950 Dust deposition from chimney stacks *Proc. Inst. Mech. Eng. (Lond.)* **162** 355–68

Bosanquet C H and Pearson J L 1936 The spread of smoke and gases from chimneys *Trans. Faraday Soc.* **32** 1249–64

Bouhuys A, Beck G J and Schoenberg J B 1978 Do present levels of air pollution outdoors affect respiratory health? *Nature* **276** 466–71

Bowne N E 1974 Diffusion rates *J. Air Pollut. Control Assoc.* **24** 832–5

Braun R C and Wilson M J G 1970 The removal of atmospheric sulphur by building stones *Atmos. Environ* **4** 371–8

Brice K A and Derwent R G 1978 Emissions inventory for hydrocarbons in the United Kingdom *Atmos. Environ.* **12** 2045–54

Briggs G A 1969 *Plume Rise* AEC Critical Review Series TID-25075 (Oak Ridge, Tennessee: US Atomic Energy Commission)

—— 1971 Some recent analyses of plume rise observations in *Proc. 2nd Int. Clean Air Conf. Washington, DC, 1970* ed H M Englund and W T Beery (New York: Academic) pp 1029–32

—— 1975 Plume rise predictions in *Lectures on Air Pollution and Environmental Impact Analyses* (Boston: American Meteorological Society) pp 59–111

Brimblecombe P 1975 Industrial air pollution in thirteenth-century Britain *Weather* **30** 388–96

—— 1977 London air pollution, 1500–1900 *Atmos. Environ.* **11** 1157–62

—— 1978 Air pollution in industrializing England *J. Air Pollut. Control Assoc.* **28** 115–18

—— 1982 Trends in the deposition of sulphate and total solids in London *Science of the Total Environment* **22** 97–103

—— 1982 An anecdotal history of air pollution in England *Environ. Educ. Inf.* **2** 97–106

Brimblecombe P and Ogden C 1977 Air pollution in art and literature *Weather* **32** 285–91

Broadbent D H 1973 Pollution as an international problem *Lecture at Sheffield University* (quoted in Science Research Council (1976))

Budiansky S 1980 Dispersion modelling *Environ. Sci. Technol.* **14** 370–4

Buell C E 1975 *Objective Procedures for Optimum Location of Air Pollution Observation Stations* EPA–650/4–75–005 (Research Triangle Park, North Carolina: USEPA)

Carroll J D, Craxford S R, Newall H E and Weatherley M–L P M 1960 Trends in the pollution of the air of Great Britain by smoke and sulphur dioxide 1952–59 *Proc. Clean Air Conf. Harrogate, 1960* (National Society for Clean Air)

Chan T L and Lawson D R 1981 Characteristics of cascade impactors in size determination of diesel particles *Atmos. Environ.* **15** 1273–9

Chan W H, Ro C U, Lusis M A and Vet R J 1982 Impact of the INCO nickel smelter emissions on precipitation quality in the Sudbury area *Atmos. Environ.* **16** 801–14

Chandler T J and Elsom D M Meteorological controls upon ground level concentrations of smoke and sulphur dioxide in two urban areas of the UK *Warren Spring Laboratory LR 245 (AP)*

Chimney Heights Third Edition of the 1956 Clean Air Act Memorandum (London: HMSO)

Clarke A G, Gascoigne M, Henderson-Sellers A and Williams A 1978 Modelling air pollution in Leeds (UK) *Int. J. Environ. Stud.* **12** 121–32

Clean Air Act 1956 (London: HMSO)

Clean Air Act 1968 (London: HMSO)

Coulter R L and Underwood K H 1980 Some turbulence and diffusion parameter estimates within cooling tower plumes derived from SODAR data *J. Appl. Meteorol.* **19** 1395–404

Courbon P 1983 Problèmes posés par la mesure des émissions d'amiante dans les cheminées *Pollution Atmospherique* **97** 10–13

Craxford S R, Weatherley M-L P M and Gooriah B D 1972 The United Kingdom, a summary *National Survey of Air Pollution 1961–71* vol 1, part 2 (London: HMSO)

Csanady G T 1973 *Turbulent Diffusion in the Environment* (2nd edn 1980) (Dordrecht: Reidel)

Cullis C F and Hirschler M M 1980 Atmospheric sulphur: natural and man-made sources *Atmos. Environ.* **14** 1263–78

Daisey J M and Lioy P J 1981 Transport of PAH into New York City *J. Air Pollut. Control Assoc.* **31** 567–9, 614

Davison D S, Fortems C C and Grandia K L 1977 The application of turbulence measurements to dispersion of a large industrial effluent plume in *Proc. Jt. Conf. on Applications of Air Pollut. Meteorol., Salt Lake City, Utah, November 29-December 2, 1977* (Boston: American Meteorological Society) pp 103–10

Department of Energy Statistical Bulletin *Energy Trends* issued monthly

Department of the Environment 1974, 1975 *Odours Report of the Working Party on the Suppression of Odours from Offensive and Selected Other Trades,* Parts 1 and 2 (Stevenage, Herts: Warren Spring Laboratory)

Derwent R G and Hov Ø 1982 The potential for secondary pollutant formation in the atmospheric boundary layer in a high pressure situation over England *Atmos. Environ.* **16** 655–65

Derwent R G and Stewart H N M 1973 Air pollution from the oxides of nitrogen in the United Kingdom *Atmos. Environ.* **7** 385–401

Diamant R M E 1974 *The Prevention of Pollution* (London: Pitman)

Direttive sull'altezza minima dei camini/Richtlinien über die Mindesthöhe von Kaminen 2 July 1980, Département Fédéral de l'Intérieur, Switzerland

Dittberner G J 1978 Climatic change: volcanoes, man-made pollution, and carbon dioxide *IEEE Trans. on Geoscience Electronics* **GE-16** 50–61

Doan M H and East C 1977 A proposed air quality index for urban areas *Water, Air, and Soil Pollut.* **8** 441–51

Dobbins R A 1979 *Atmospheric Motion and Air Pollution* (New York: Wiley)

Drufuca G, Giugliano M and Torlaschi E 1980 SO_2 dosages in urban areas *Atmos. Environ.* **14** 11–17

Durham J L, Overton J H Jr and Aneja V P 1981 Influence of gaseous nitric acid on sulfate production and acidity in rain *Atmos. Environ.* **15** 1059–68

Ellis H M and Greenway A R 1981 The prevention of significant deterioration of air quality. Summary of the final Federal Regulation (August 7 1981) *J. Air Pollut. Control Assoc.* **31** 136–8

Ellis H M, Liu P C and Runyon C 1980 Comparison of predicted and measured concentrations for 58 alternative models of plume transport in complex terrain *J. Air Pollut. Control. Assoc.* **30** 670–5

Elsom D M 1979 Air pollution episode in Greater Manchester *Weather* **34** 277–86

Esche M 1983 Stack gas desulfurization without reheating in *Proc. VIth World Congress on Air Quality* vol 3 (Paris: IUAPPA) pp 479–85

Evans J S and Cooper D W 1980 An inventory of particulate emissions from open sources *J. Air Pollut. Control Assoc.* **30** 1298–303

Evelyn J *Fumifugium* (reprinted by the National Society for Clean Air, 1960)

Faoro R B and Manning J A 1981 Trends in benzo(a)pyrene 1966–1977 *J. Air Pollut. Control Assoc.* **31** 62–4

Federal Register 1971 *National primary and secondary ambient air quality standards* vol 36, no 84, Part II

Fox D G 1970 Forced plume in a stratified fluid *J. Geophys. Res.* **75** 6818–35

Frye R S and Ayers K C 1981 Air permits for new and modified sources: the significance of June 8, 1981 *J. Air Pollut. Control Assoc.* **31** 397–400

Galloway J N and Likens G E 1981 Acid precipitation: the importance of nitric acid *Atmos. Environ.* **15** 1081–5

Galloway J N and Whelpdale D M 1980 An atmospheric sulfur budget for eastern North America *Atmos. Environ.* **14** 409–18

Gardner M J, Winter P D and Acheson E D 1982 Variations in cancer mortality among local authority areas in England and Wales: relations with environmental factors and search for causes *Br. Med. J.* **284** 784–7

Garner J F and Crow R K 1976 *Clean Air—Law and Practice* 4th edn (London: Shaw and Sons)

Garnett A 1980 Recent trends in sulphur dioxide air pollution in the Sheffield urban region *Atmos. Environ.* **1** 787–96

Garnett A, Read P and Finch D 1976 The use of conversion factors in air pollution studies: sulphur dioxide ppm—μg m^{-3} *Atmos. Environ.* **10** 325–8

Gibson J 1978 The 1977 Robens Coal Science Lecture: The constitution of coal and its relevance to coal conversion processes *J. Inst. Fuel* **51** 67–81

Gifford F A Jr 1960 Atmospheric dispersion *Nuclear Safety* **1** 56–9

—— 1972 Atmospheric transport and dispersion over cities *Nuclear Safety* **13** 391–402

—— 1975 Atmospheric dispersion model for environmental pollution applications in *Lectures on Air Pollution and Environmental Impact Analyses* (Boston: American Meteorological Society) pp 35–58

Gilmore G N 1979 *A Modern Approach to Comprehensive Chemistry* 2nd edn (Cheltenham: Stanley Thorne)

Gorse R A Jr and Norbreck J M 1981 CO emission rates for in-use gasoline and diesel vehicles *J. Air Pollut. Control Assoc.* **31** 1094–6

Graedel T E 1977 The wind boxplot: an improved wind rose *J. Appl. Meteorol.* **16** 448–50

Grandjean P and Nielsen T 1979 Organolead compounds: environmental health aspect *Resid. Rev.* **72** 97–148

Green A E S, Singhal R P and Venkateswar R 1980 Analytic extensions of the Gaussian plume model *J. Air Pollut. Control Assoc.* **30** 773–6

Greene R 1977 Utilities scrub out SO_2 *Chem. Eng.* **23** 101–3 (May)

Hales J M, Wolf M A and Dana M T 1973 A linear model for predicting the washout of pollutant gases from industrial plumes *A. I. Ch. E. J.* **19** 292–7

van Haluwyn Ch and Lerond M 1983 Lichens: vegetaux de la pollution atmospherique in *Proc. VIth World Congress on Air Quality* vol 2 (Paris: IUAPPA) pp 469–74

Hamilton P M 1969 Use of LIDAR in the study of chimney plumes *Phil. Trans. R. Soc.* A. **265** 153–72

Hampel C A and Hawley G G 1976 *Glossary of Chemical Terms* (New York: Van Nostrand)

Hanna S R, Briggs G A, Deardorff J, Egan B A, Gifford F A and Pasquill F 1977 AMS workshop on stability classification schemes and sigma curves—summary of recommendations *Bull. Am. Meteorol Soc.* **58** 1305–9

Hansen J, Johnson D, Lacis A, Lebedeff S, Lee P, Rind D and Russell G 1981 Climate impact of increasing atmospheric carbon dioxide *Science* **213** 957–66

Harkort W and Brose G 1983 Olfactometric measurements and dispersion calculations in *Proc. VIth World Congress on Air Quality* vol 2 (Paris: IUAPPA) pp 431–4

Hawkins J E and Nonhebel G 1955 Chimneys and dispersal of smoke *J. Inst. Fuel* **28** 530–45

Hawksworth D L and Rose F 1970 Qualitative scale for estimating sulphur dioxide air pollution in England and Wales using epiphytic lichens *Nature* **227** 145–8

Health and Safety Executive 1984 Occupational exposure limits *Guidance Note EH40/84* (London: HMSO)

Henderson-Sellers A 1980 Air pollutant levels: difficulties of assessment of area values and trends *Water, Air, and Soil Pollut.* **13** 173–86

Henderson-Sellers A and Seaward M R D 1979 Monitoring lichen reinvasion of ameliorating environments *Environ. Pollut.* **19** 207–15

Henderson-Sellers B 1980 The behaviour of marginally buoyant plumes in an urban environment *Ecol. Model.* **9** 43–56

—— 1981, Shape constants for plume models *Boundary-Layer Meteorol.* **21** 105–14

Henderson-Sellers B and Newsum D A 1980 Energy resources—are we planning for the future?—A review *Int. J. Energy Res.* **4** 225–33

Hirschler M M 1981 Man's emission of carbon dioxide into the atmosphere *Atmos. Environ.* **15** 719–27

HMSO 1932 *The measures which have been taken in this country and in others to obviate the emission of soot, ash, grit and gritty particles from the chimneys of electric power stations* by a committee appointed by the Electricity Commissioners, London (London: HMSO)

—— 1954 Mortality and morbidity during the London fog of December 1952 *Ministry of Health Report No. 95* (London: HMSO)

—— 1981 *Digest of Environmental Pollution and Water Statistics* No 4 Additional Tables (London: HMSO)

—— 1982 *Digest of Environmental Pollution and Water Statistics* No 5 (London: HMSO)

Hodges L 1973 *Environmental Pollution* (2nd edn 1977) (New York: Holt, Rinehart and Winston)

Horsman D C, Roberts T M and Bradshaw A D 1978 Evolution of sulphur dioxide tolerance in perennial rye grass *Nature* **276** 493–4

Hov Ø and Derwent R G 1981 Sensitivity studies of the effects of model formulation on the evaluation of control strategies for photochemical air pollution formation in the United Kingdom *J. Air Pollut. Control Assoc.* **31** 1260–7

Hunt J M 1981 The origin of petroleum *Oceanus* **24** 53–7

Irwin J S 1979 Estimating plume dispersion—a recommended generalized scheme in *Proc. 4th Symp. on Turbulence, Diffusion and Air Pollution, Reno, Nevada, January 1979* (Boston: American Meteorological Society) pp 62–9

—— 1983 Estimating plume dispersion—a comparison of several sigma schemes *J. Clim. Appl. Meteorol.* **22** 92–114

Jarrault P 1978 *Limitation des émissions de pollutants et qualité de l'air, valeurs reglementaires dans les principaux pays industrialises, specifications envigueur en 1978* (Paris: Institut Français de l'Energie)

Joyce C 1980 Industrial China's expensive dirt *New Scientist* **87** 772–5

Kerekes J, Howell G, Beauchamp S and Pollock T 1982 Characterization of three lake basins sensitive to acid precipitation in central Nova Scotia (June 1979 to May 1980) *Int. Revue ges. Hydrobiol.* **67** 679–94

Koschmieder H 1924 Theorie der horizontalen Sichtweile *Beitr. Physik Freien Atmosphäre* **12** 33, 171

Kumar A 1979 Air quality at the tar sands *Env. Sci. Technol.* **13** 650–4

Leach G, Lewis C, Romig F, Foley G and van Baren A A 1979 *A Low Energy Strategy for the United Kingdom* (London: IIED Science Reviews)

Ledbetter J P 1972 *Air Pollution* parts A and B (New York: Marcel Dekker)

Lee R F, Tikvart J A, Dicke J L and Fisher R W 1978 The effect of revised dispersion parameters on concentration estimates in *Proc. 4th Symp. on Turbulence, Diffusion and Air Pollution, Reno, Nevada, January 1979* (Boston: American Meteorological Society) pp 70–4

Lee S D (ed) 1980 *Nitrogen Oxides and Their Effects on Health* (Ann Arbor: Ann Arbor Science Publishers)

Leonardos G, Kendall D and Barnard N 1969 Odor threshold determination of 53 odorant chemicals *J. Air Pollut. Control Assoc.* **19** 91–5

Levine J S 1984 Water and the photochemistry of the troposphere in *Satellite Sensing of a Cloudy Atmosphere: Observing the Third Planet* (ed A Henderson-Sellers) (London: Taylor and Francis)

Levine J S and Allario F 1982 The global troposphere: biogeochemical cycles, chemistry and remote sensing *Env. Monit. Assess.* 1 263–306

Likens G E 1976 Acid precipitation *Chem. and Eng. News* 54 29–44

Likens G E and Butler T J 1981 Recent acidification of precipitation in North America *Atmos. Environ.* 15 1103–9

Littlejohn J L, Shaver D B and Malm W C 1981 The inadequacy of PSD increments to protect visibility in Class 1 areas *J. Air Pollut. Control Assoc.* 31 879–80

Liu B Y H and Pui D Y H 1981 Aerosol sampling inlets and inhalable particles *Atmos. Environ.* 15 589–600

Lorenz E N 1975 Climatic predictability in *The Physical Basis of Climate and Climate Modeling* GARP Publication Series No 16 (WMO/ICSU) 132–7

Lott R A 1982 Terrain-induced downwash effects on ground level SO_2 concentrations *Atmos. Environ.* 16 635–42

Lovelock J E 1979 *The Gaia Hypothesis* (Oxford: Oxford University Press)

Lucas D H 1967 Application and evaluation of results of the Tilbury rise and dispersion experiment *Atmos. Environ.* 1 421–4

Ludwick J D, Weber D B, Olsen K B and Garcia S R 1980 Air quality measurements in the coal fired power plant environment of Colstrip, Montana *Atmos. Environ.* 14 523–32

Luten D B 1974 United States requirements in *Energy, The Environment and Human Health* (ed J Finkel) (Acton, Mass.: Publishing Sciences Group)

Mabey R 1974 *The Pollution Handbook* (Harmondsworth: Penguin Education)

Mackintosh B C and Coleman C R 1983 Oxides of nitrogen emissions from utility boilers in *Proc. VIth World Congress on Air Quality* vol 3 (Paris: IUPPA) pp 465–72

McMullan J T, Morgan R and Murray R B 1977 *Energy Resources* (London: Edward Arnold)

Martin L R, Damschen D E and Judeikis H S 1981 The reactions of nitrogen oxides with SO_2 in aqueous aerosols *Atmos. Environ.* 15 191–5

Meetham A R, Bottom D W, Cayton S, Henderson-Sellers A and Chambers D 1981 *Atmospheric Pollution. Its History, Origins and Prevention* 4th edn (Oxford: Pergamon)

Melia R J W, Florey C du V, Altman D G and Swan A V 1977 Association between gas cooking and respiratory disease in children *Br. Med. J.* 2 149–52

Melia R J W, Florey C duV, Sittampalam Y M and Watkins C J 1983 The relation between respiratory illness in infants and gas cooking in the UK: a preliminary report in *Proc. VIth World Congress on Air Quality* vol 2 (Paris: IUAPPA) pp 263–9

Middleton D R, Butler J D and Colwill D M 1979 Gaussian plume dispersion model applicable to a complex motorway interchange *Atmos. Environ.* 13 1039–49

Miller C W 1978 An examination of Gaussian plume dispersion parameters for rough terrain *Atmos. Environ.* 12 1359–64

Miller F J, Gardner D E, Graham J A, Lee R E Jr, Wilson W E and Bachmann J D 1979 Size considerations for establishing a standard for inhalable particles *J. Air Pollut. Control Assoc.* 29 610–15

Mitchell A E Jr 1982 A comparison of short-term dispersion estimates resulting from various atmospheric stability classification methods *Atmos. Environ.* **16** 765–73

Morton B R 1959 The ascent of turbulent forced plumes in a calm atmosphere *Int. J. Air Pollut.* **1** 184–97

—— 1971 The choice of conservation equations for plume models *J. Geophys. Res.* **76** 7409–16

Morton B R, Taylor G I and Turner J S 1956 Turbulent gravitational convection from maintained and instantaneous sources *Proc. R. Soc.* A **234** 1–23

Munn R E 1981 *The Design of Air Quality Monitoring Networks* (London: Macmillan)

National Coal Board 1980 Fluidised bed combustion of coal *NCB Report*

—— undated *Coal for Industry*

National Society for Clean Air *Annual Reports*

National Survey of Air Pollution *Annual Reports*

National Survey of Smoke and Sulphur Dioxide 1966 *Instruction Manual* (Stevenage, Herts: Warren Spring Laboratory)

Needleman H L, Gunnoe C, Levitan A, Read R, Peresie H, Maher C and Barrett P 1979 Deficits in psychological and classroom performance of children with elevated dentine lead levels *New Engl. J. Med.* **300** 689–95

Neumann G and Halbritter G 1980 Sensitivity analysis of the Gaussian plume model in *Atmospheric Pollution 1980* ed M M Benarie (Amsterdam: Elsevier) pp 57–62

New Scientist 1981 *CFCs* **91** 212–14

Noll K E and Miller T L 1977 *Air Monitoring Survey Design* (Ann Arbor: Ann Arbor Science Publishers)

Noll K E, Miller T L, Norco J E and Raufer R K 1977 An objective air monitoring site selection methodology for large point sources *Atmos. Environ.* **11** 1051–9

Nonhebel G 1960 Recommendations on heights for new industrial chimneys *J. Inst. Fuel* **33** 479

—— 1975 Best practicable means and presumptive limits—British definitions *Atmos. Environ* **9** 709–15

—— 1981 Industrial air pollution: British progress—a review *Atmos. Environ.* **15** 213–19

OECD 1964 *Methods of measuring air pollution* (Paris: OECD)

—— 1977 *Report on programme on long range transport of air pollutants* (Paris: OECD)

Ooms G 1972 A new method for all calculations of the plume path of gases emitted by a stack *Atmos. Environ.* **6** 899–909

Open University 1975 PT272 Units 13–14 *Air Pollution*

—— 1975 PT272 Unit 15 *Air Pollution Control*

Parker A (ed) 1978 *Industrial Air Pollution Handbook* (New York: McGraw-Hill)

Parker H W 1977 *Air Pollution* (New York: Prentice-Hall)

Pashel G E and Egner D R 1981 A comparison of ambient suspended particulate matter concentrations as measured by the British smoke sampler and the high volume sampler at 16 sites in the United States *Atmos. Environ.* **15** 919–27

Pasquill F 1971 Atmospheric dispersion of pollution *Q. J. R. Meteorol. Soc.* **97** 369–95

—— 1974 *Atmospheric Diffusion* 2nd edn (Chichester: Ellis Horwood)

—— 1976 *Atmospheric Dispersion Parameters in Gaussian Plume Modelling, Part*

2, Possible Requirements for Change in the Turner Workbook Values EPA–600/4–76–030b (Washington, DC: US Govt. Printing Office)

—— 1979 Atmospheric dispersion modelling *J. Air Pollut. Control. Assoc.* **29** 117–19

Paterson M P and Benjamin S F 1975 Better than a wind rose *Atmos. Environ.* **9** 537–42

Perkins H C 1974 *Air Pollution* (New York: McGraw-Hill)

Posthumus A 1983 General philosophy for the use of plants as indicators and accumulators of air pollutants and as bio-monitors of their effects in *Proc. VIth World Congress on Air Quality* vol 2 (Paris: IUAPPA) pp 555–62

Priestley C H B 1959 *Turbulent Transfer in the Lower Atmosphere* (Chicago: University of Chicago Press)

Rapolla A, Keller G V and Moore D J (eds) 1980 *Geophysical Aspects of the Energy Problem* (Amsterdam: Elsevier)

Rashidi M and Massoudi M S 1980 A study of the relationship of street level carbon monoxide concentrations to traffic parameters *Atmos. Environ.* **14** 27–32

Repace J L and Lowrey A H 1980 Indoor air pollution, tobacco smoke, and public health *Science* **208** 464–72

Richardson D H S and Nieboer E 1981 Lichens and pollution monitoring *Endeavour* **5** 127–33

Rittmann B E 1982 Application of two-thirds law to plume rise from industrial-sized sources *Atmos. Environ.* **16** 2575–9

Ross R D 1972 *Air Pollution and Industry* (New York: Van Nostrand)

Rowat D W and Meisen A 1983 Siting of ambient monitoring stations in *Proc. VIth World Congress on Air Quality* vol 3 (Paris: IUAPPA) pp 127–33

Rubin E S 1981 Air pollution constraints on increased coal use by industry *J. Air Pollut. Control Assoc.* **31** 349–60

Rycroft M J 1982 Analysing atmospheric carbon dioxide levels *Nature* **295** 190–1

Schidlowski M 1980 The atmosphere in *The Handbook of Environmental Chemistry* vol 1 part A (ed O Hutzinger) (Berlin: Springer)

Schiermeier F A, Wilson W E, Pooler F, Ching J K S and Clarke J F 1979 Sulfur transport and transformation in the environment (STATE): a major EPA research program *Bull. Am. Meteorol. Soc.* **60** 1303–12

Science Research Council 1976 *Combustion-Generated Pollution* (London: Science Research Council)

Scorer R S 1968 *Air Pollution* (Oxford: Pergamon)

—— 1973 *Pollution in the Air* (London: Routledge and Kegan Paul)

—— 1978 *Environmental Aerodynamics* (Chichester: Ellis Horwood)

Scott J A 1963 The London fog of December 1962 *Med. Officer* **109** 250–2

Sedefian L and Bennett E 1980 A comparison of turbulence classification schemes *Atmos. Environ.* **14** 741–50

Seinfeld J H 1975 *Air Pollution. Physical and Chemical Fundamentals* (New York: McGraw-Hill)

—— 1980 *Lectures in Atmospheric Chemistry* American Institute of Chemical Engineers Monograph Series 12 vol 76

Semb A 1978 Sulphur emissions in Europe *Atmos. Environ.* **12** 455–60

Sittig M 1975 *Environmental Sources and Emissions Handbook* (Park Ridge, New Jersey: Noyes Data Corporation)

Skinner D G 1971 *The fluidised combustion of coal* M and B Monographs, Chemical Engineering (London: Mills and Boon)

Slawson P R and Csanady G T 1967 On the mean path of buoyant bent over chimney plumes *J. Fluid Mech.* **28** 311–27

—— 1971 The effect of atmospheric conditions on plume rise *J. Fluid Mech.* **47** 33–49

Smith M E (ed) 1968 *Recommended Guide for the Prediction of the Dispersion of Airborne Effluents* (New York: American Society of Mechanical Engineers)

Snyder W H and Lawson R E Jr 1976 Determination of a necessary height for a stack close to a building—a wind tunnel study *Atmos. Environ.* **10** 683–91

Stalker W W, Dickerson R C and Kramer G D 1962 Sampling station and time requirements for urban air pollution surveys, Part IV *J. Air Pollut. Control Assoc.* **12** 361–75

Steed J M, Owens A J, Miller C, Filkin D L and Jesson J P 1982 Two-dimensional modelling of potential ozone perturbation by chlorofluorocarbons *Nature* **295** 308–11

Stephens R 1981 Human exposure to lead from motor vehicle emissions *Int. J. Env. Stud.* **17** 73–83

Stern A C (ed) 1976 *Air Pollution* 3rd edn 5 vols (New York: Academic)

—— 1982 History of air pollution legislation in the United States *J. Air Pollut. Control Assoc.* **32** 44–61

Stern A C, Wohlers H C, Boubel R W and Lowry W P 1973 *Fundamentals of Air Pollution* (New York: Academic)

Stock S 1980 The perils of second-hand smoking *New Scientist* **88** 10–13

Strauss W 1978 *Air Pollution Control* 3 vols (New York: Wiley)

Sutton O G 1932 A theory of eddy diffusion in the atmosphere *Proc. R. Soc.* A **135** 143–65

—— 1947 The theoretical distribution of airborne pollution from factory chimneys *Q. J. R. Meteorol. Soc.* **73** 426–36

—— 1953 *Micrometeorology* (New York: McGraw-Hill)

Sze N D and Ko M K W 1980 Photochemistry of COS, CS_2, CH_3SCH_3 and H_2S: implications for the atmospheric sulphur cycle *Atmos. Environ.* **14** 1223–39

—— 1981 The effects of the rate for $OH + HNO_3$ and HO_2NO_2 photolysis on stratospheric chemistry *Atmos. Environ.* **15** 1301–7

Taylor G I 1921 Diffusion by continuous movements *Proc. Math. Soc. Lond.* **20** 196–212

Temple P J, McLaughlin D L, Linzon S N and Wills R 1981 Moss bags as monitors of atmospheric deposition *J. Air Pollut. Control Assoc.* **31** 668–70

Tercier P 1981 Revue et analyse des coefficients de dispersion atmospherique exprimes sous la forme ax^b *Working Report of the Swiss Meteorological Institute* No 105

Thain W 1980 *Monitoring Toxic Gases in the Atmosphere for Hygiene and Pollution Control* (Oxford: Pergamon)

Torrens I M 1983 Coal: environmental issues, remedies and their costs in *Proc. VIth World Congress on Air Quality* vol 2 (Paris: IUAPPA) pp 283–90

Tout D G 1973 Manchester sunshine *Weather* **28** 164–6

Turner D B 1970 Workbook of Atmospheric Dispersion Estimates *US Public Health*

Service Publication 999–AP–26 (revised edition)(Washington, DC: US Govt. Printing Office)

—— 1979 Atmospheric dispersion modelling. A critical review *J. Air Pollut. Control Assoc.* **29** 502–19

20 Years of Air Pollution Control Manchester Area Council for Clean Air and Noise Control

Van Egmond N D and Onderdelinden D 1981 Objective analysis of air pollution monitoring network data: spatial interpolation and network density *Atmos. Environ.* **15** 1035–45

Vick C M and Bevan R 1976 Lichens and tar spot fungus (*Rhytisma acerinum*) as indicators of sulphur dioxide pollution on Merseyside *Environ. Pollut.* **11** 203–16

Voldner E C, Shah Y and Whelpdale D M 1980 A preliminary Canadian emissions inventory for sulfur and nitrogen oxides *Atmos. Environ.* **14** 419–28

Waller R 1983 The influence of urban air pollution on the development of chronic respiratory disease in *Proc. VIth World Congress on Air Quality* vol 2 (Paris: IUAPPA) pp 51–7

Wang L K and Pereira N C (eds) 1979 *Handbook of Environmental Engineering, Vol 1, Air and Noise Pollution Control* (Clifton, NJ: Humana Press)

Weatherley M-L P M 1977 Fuel consumption, and smoke and sulphur dioxide emissions, in the United Kingdom up to 1976 *Report LR 258(AP)* (Stevenage, Herts: Warren Spring Laboratory)

—— 1979 The National Survey of smoke and sulphur dioxide – quality control and the air sampling arrangements *Report LR 308 (AP)* (Stevenage, Herts: Warren Spring Laboratory)

Weber E 1981 Air pollution control strategy in the Federal Republic of Germany *J. Air Pollut. Control Assoc.* **31** 24–30

Weil J C and Hoult D P 1973 A correlation of ground level concentrations of sulfur dioxide downwind of the Keystone stacks *Atmos. Environ.* **7** 707–21

Whaley H and Lee G K 1982 The behaviour of buoyant plumes in neutral and stable conditions in Canada *Atmos. Environ.* **16** 2555–64

Whitby K T 1977 The physical characteristics of sulfur aerosols *Paper prepared for the Int. Symp. on Sulphur in the Atmosphere, Sept 7–14 1977, Dubrovnik, Yugoslavia*

Wilkins E T 1954 Air pollution aspects of the London fog of December 1952 *Q. J. R. Meteorol. Soc.* **80** 267–71

Williamson S J 1973 *Fundamentals of Air Pollution* (Reading, Mass.: Addison-Wesley)

Wilson D G 1978 Alternative automobile engines *Sci. Am.* **239** (1) 27–37

Wood C M and Lee N 1976 Cities and pollution *Int. J. Environ. Stud.* **8** 293–300

Wood C M, Lee N, Laker J A and Saunders P J W 1974 *The Geography of Pollution* (Manchester: Manchester University Press)

Worth J B and Ripperton L A 1980 Rural ozone—sources and transport in *Advances in Environmental Science and Engineering* (ed J R Pfafflin and E N Ziegler) vol 3 (New York: Gordon and Breach) pp 150–170

Yule W, Lansdown R, Millar I B and Urbanowl M A 1981 The relationship between blood lead concentrations, intelligence and attainment in a school population—a pilot study *Develop. Med. Child Neurol.* **23** 567–76

Index